ANTIBACTERIAL AGENTS

Edited by **Ranjith N. Kumavath**

Antibacterial Agents
http://dx.doi.org/10.5772/65630
Edited by Ranjith N. Kumavath

Contributors

Laurence Walsh, Patricia Wright, Marija Vukomanovic, Mario Kurtjak, Nemanja Anicic, Horea Vladi Matei, Mihaela Laura Vica, Costel Vasile Siserman, Jessica M. Rosenholm, Didem Sen Karaman, Suvi Manner, Adyary Fallarero, Monica Periolatto, Franco Ferrero, Claudia Vineis, Alessio Varesano, Giuseppe Gozzelino, Hamid Ullah, Saqib Ali

Published by InTech

Janeza Trdine 9, 51000 Rijeka, Croatia

Publishing Process Manager Dajana Pemac
Technical Editor SPi Global
Cover InTech Design team

First published May, 2017
Printed in Croatia
Legal deposit, Croatia: National and University Library in Zagreb

Additional hard copies can be obtained from orders@intechopen.com

Antibacterial Agents, Edited by Ranjith N. Kumavath
p. cm.
Print ISBN 978-953-51-3199-1
Online ISBN 978-953-51-3200-4

For EU product safety concerns:
IN TECH d.o.o., Prolaz Marije Krucifikse Kozulić 3, 51000 Rijeka, Croatia,
info@intechopen.com or visit our website at intechopen.com.

Meet the editor

Dr. Ranjith Kumavath, MSc, PhD, FASB, was born in 1979 to a lovely Rajput Indian couple and grew up in south India. He received his PhD degree from the University of Hyderabad. Before joining CUK in 2011, he worked as a postdoctoral research fellow at the University of Hyderabad, Singapore and the USA. He has been awarded as Young Scientist by DST, Govt. of India; BioAsia Young Scientist by govt. of Andhra Pradesh; Fellow of the Society of Applied Biotechnology by SAB; Fellow of Young Investigator by India Biosciences and Mahatma Gandhi Gold Medal Award 2014; and APJ Abdul Kalam Award by GEPRA, India. Currently, he is operating four major research projects funded by UGC-BSR, SERB-DST and SERB, Govt. of India. He has published 2 books, 6 chapters and 23 research publications in leading international journals, and he has made 18 national and 16 international presentations. He has contributed in discovery of four novel enzymes to IUBMB. He has collaborations with UoH, IIOAB, Cancer Science Institute of Singapore and San Diego State University, USA. His main research interest areas are cancer genomics, microbial genetics and metagenomics, human infective diseases and computational drug designing. His research group is being dedicated towards developing anti-cancer agents from microbial recourses for target-based drug discovery.

Contents

Preface

To cure infectious diseases, many researchers discovered antibacterial agents that are considered as probably the most promising chemotherapeutic agents. The pathogenic microbes cause diseases and infections when they get into the body and begin to reproduce and crowd out healthy bacteria and/or to grow into tissues that are normally sterile. Keeping in mind the resistance phenomenon developing against antibacterial agents, new drugs are frequently entering into the market along with the existing drugs. The antibacterial agents can be discussed in five major classes, i.e. classification based on the type of action, source, spectrum of activity, chemical structure and function.

Resistance of bacteria to antibiotics is an urgent problem of humanity, which leads us to the lack of therapy for serious bacterial infections. Development of new antibiotics has almost ceased in the last decades—even when a new antibiotic is launched, very soon the resistance of bacteria appears. There is a long list of applications where antimicrobial protection is requested to achieve effective treatment. The loss of effective antibiotic treatments will not only cripple the ability to fight routine infectious diseases but will also undermine treatment of infectious complications in patients with other diseases. For example, sexually transmitted diseases (STD) are among the most common infections worldwide predominantly in the developing/underdeveloped countries, and also endodontic treatment, for a multispecies bacterial and fungal infection that is present in a place that is inaccessible to the host immune system, offers physical protection from applied topical agents.

However, they will be irrigating various deficits in performance, which is why many clinicians recommended alternative approaches to using antimicrobial substances. Industrial textiles exposed as awnings, screens, tents; upholstery used in large public areas such as hospitals, hotels and stations; fabrics for transports; protective clothing and personal protective equipment; bed sheets and blankets; textiles left wet between processing steps; intimate apparel, underwear, socks and sportswear, disinfection of air and water for white rooms, hospitals and operating theatres, food and pharma industries, water depuration, drinkable water supplying and air conditioning systems. The majority of bioagents demonstrate on antibiotics for treatment of a wide range of diseases in human sectors. However, the misuse and mishandling of drugs lead to microbial, particularly bacterial, resistance as well as result in the difficulty of treating microbial diseases. Hence, the proposed book will give more precise information on novel antibacterial compound(s).

Dr. Ranjith N. Kumavath, MSc, PhD, FSAB
Sr. Assistant Professor and Principal Investigator
Department of Genomic Science
School of Biological Sciences
Central University of Kerala, India

Chapter 1

Classification of Anti-Bacterial Agents and Their Functions

Hamid Ullah and Saqib Ali

Additional information is available at the end of the chapter

http://dx.doi.org/10.5772/intechopen.68695

Abstract

Bacteria that cause bacterial infections and disease are called pathogenic bacteria. They cause diseases and infections when they get into the body and begin to reproduce and crowd out healthy bacteria or to grow into tissues that are normally sterile. To cure infectious diseases, researchers discovered antibacterial agents, which are considered to be the most promising chemotherapeutic agents. Keeping in mind the resistance phenomenon developing against antibacterial agents, new drugs are frequently entering into the market along with the existing drugs. In this chapter, we discussed a revised classification and function of the antibacterial agent based on a literature survey. The antibacterial agents can be classified into five major groups, i.e. type of action, source, spectrum of activity, chemical structure, and function.

Keywords: anti-bacterial agents, classification, functions

1. Introduction

Bacteria are simple one-celled organism, which were first identified in the 1670s by van Leeuwenhoek. Latter in the nineteenth century, concepts have been developed that there is the strongest correlation between bacteria and diseases. Such considerations attracted interest of the researchers not only to answer some mysterious questions about infectious diseases, but also to find a substance that could kill, inhibit, or at least slow down the growth of such disease-causing bacteria. These efforts led to the revolutionary discovery of the antibacterial agent "penicillin" in 1928 from *Penicillium notatum* by Sir Alexander Fleming. The discovery unlocked the field of microbial natural products and so new agents were continually added, such as newly introduced daptomycin, tigecycline, linezolid, and so on. Gradually, due to various issues arising during the use of antibacterial agents, such as the resistance phenomenon, an enormous increase in the number and types (e.g., structurally different and agent with a slightly different pattern of activity) of the newly added antibacterial agents has been observed, which made it necessary to review and compile the existing classification and functions of almost all the antibacterial agents. It is aimed that this approach will be equally helpful for researchers, clinicians, and academicians.

2. Classification

Infectious diseases are the major causes of human sickness and death. To overcome such health care issues, antibiotics proved to be promising agents ever since they were introduced in the 1940s. Antibacterials, which are a subclass of antibiotics, have been classified earlier in several ways; however, to make it more easily understandable, we can classify antibacterial agents into five groups: type of action, source, spectrum of activity, chemical structure, and function [1].

2.1. Classification based on type of action

Generally, antibacterials can be classified on the basis of type of action: bacteriostatic and bactericidal. Antibacterials, which destroy bacteria by targeting the cell wall or cell membrane of the bacteria, are termed bactericidal and those that slow or inhibit the growth of bacteria are referred to as bacteriostatic. Actually, the inhibition phenomenon of bacteriostatic agents involves inhibition of protein synthesis or some bacterial metabolic pathways. As bacteriostatic agents just prevent the growth of the pathogenic bacteria, sometimes it is difficult to mark a clear boundary between bacteriostatic and bactericidal, especially when high concentrations of some bacteriostatic agents are used then they may work as bactericidal [2]. Some prominent examples of bacteriostatic and bactericidal antibacterials along with their mode of action are presented in **Table 1**.

A. Bacteriostatic antibacterials	**Function**
Sulphonamides	They act to inhibit folate synthesis at initial stages
Amphenicols, e.g. chloroamphenicol	Amphenicols work as protein synthesis inhibitors
Spectinomycin	It binds to the 30S ribosomal subunit, thereby interrupting protein synthesis
Trimethoprim	It disturbs the tetrahydrofolate synthesis pathway
Tigecycline; it belongs to the glycylcycline class	It is a protein synthesis inhibitor. It binds reversibly to the 30S bacterial ribosomal subunit, which blocks the binding of amino-acyl-tRNA to the acceptor site on the mRNA complex. This prohibits the incorporation of amino acids to the developing peptide chain, thereby inhibiting protein synthesis.
Erythromycin, clarithromucin and azithromycin are macrolides	They work as inhibitors of protein synthesis
Linezolid is a member of the oxazolidinone class	
Doxycycline, tetracycline, and minocycline belong to tetracyclines class	

B. **Bactericidal antibacterials**	**Function**
Penicillins, e.g. pen V, penicillin G, procaine penicillin G, benzathine penicillin G, methicillin, oxacillin, cloxacillin, dicloxacillin and flucloxacillin. They belong to β-lactams antibiotic class	
Carbapenems like imipenem, meropenem, aztreonam, ticaracillinclvulnate and piperaciin-tazobactam; these are β-lactam/β-lactamase inhibitors. Some other β-lactam inhibitors are cephalosporin, e.g. cefotaxime, ceftriaxone, ceftazidime, and cefepime	They work by interfering the synthesis of the bacterial cell wall
Gentamicin, tobramycin, and amikacin are aminoglycosides	They inhibit protein synthesis
Quinolones and flouroquinolones, such as levofloxacin, ciprofloxacin, and oxifloxacin	These block bacterial DNA replication
Vancomycine is a glycopeptide	These inhibit cell wall synthesis
Polymyxin B and colistin are polymyxins	These antibacterial disrupt cell membrane

Table 1. List of some bacteriostatic and bactericiadal antibacterials.

2.2. Classification based on source of antibacterial agents

Antibacterials are the subclass of antibiotics, which can be naturally obtained from fungal sources, semi-synthetic members which are chemically altered natural product and or synthetic. Cephalosporins, cefamycins, benzylpenicillin, and gentamicin are well-known examples of natural antibiotics/antibacterials. Natural antibiotics/antibacterials often exhibit high toxicity than synthetic antibacterials. Ampicillin and amikacin are semi-synthetic antibiotics, which were developed to show low toxicity and increase effectiveness. Synthetic antibiotics are also designed to have even greater effectiveness and less toxicity and, thus, have an advantage over the natural antibiotics that the bacteria are not exposed to the compounds until they are released. Moxifloxacin and norfloxacin are promising synthetic antibiotics [3].

2.3. Classification based on spectrum of activity

This is another way of classification of antibiotics or antibacterial agents, which is based on their target specification. In this category, the antibacterials may be either narrow or broad spectrum. The terms narrow spectrum and broad spectrum have been interpreted not specifically since their use in antibiotic history, but recently these acquired clear meanings in academic and industrial fields [4, 5]. The narrow spectrum antibacterials are considered to be those which can work on a narrow range of microorganisms, that is, they act against Gram-positive only or Gram-negative only bacteria. Unlike narrow spectrum antibacterial, the broad spectrum antibacterial affects a wide range of pathogenic bacteria, including both Gram-positive and Gram-negative bacteria. Usually, the narrow spectrum antibacterials are considered ideal antibacterials and are preferred over the broad-spectrum antibacterials. The reason is that the narrow-spectrum antibiotics do not kill as many of the normal microorganisms in the body as the broad-spectrum antibiotics and thus has less ability to cause superinfection. Also, the narrow-spectrum antibiotic will cause less resistance of the bacteria as it will deal with only specific bacteria.

Based on the spectrum of activity, both of these groups have a large and diverse library of antibacterials. **Table 2** shows all the well-known examples of these categories.

2.4. Classification based on chemical structure

Different skeleton-containing antibiotics display different therapeutic behaviour; therefore, it is an ultimate need to classify antibacterials on the basis of their chemical structure. This classification is also very important as similar structural units have similar patterns of toxicity, effectiveness, and other related properties. Usually on a structural basis, antibacterials have

Broad-spectrum antibacterials (examples)	Narrow-spectrum antibacterials (examples)
Ampicillin and its derivative amoxicillin are broad-spectrum antibacterials. Amoxicillin/clavulanic acid (common name co-amoxiclav) is an antibiotic useful for the treatment of a number of bacterial infections	β-Lactamase-sensitive, first generation include penicillin G, benzathine penicillin G, penicillin V, procaine penicillin, propicillin, pheneticillin, azidocillin, clometocillin, and penamecillin are considered in narrow-spectrum antibacterial category
Quinolones [6] such as Maxaquin (lomefloxacin), Floxin (ofloxacin), Noroxin (norfloxacin), Tequin (gatifloxacin), Cipro (ciprofloxacin), Avelox (moxifloxacin), Levaquin (levofloxacin), Factive (gemifloxacin), Cinobac (cinoxacin), NegGram (nalidixic acid), Trovan (trovafloxacin), and Zagam (sparfloxacin) are considered as broad-spectrum antibacterials	β-Lactamase-resistant, Ist generation include; Cloxacillin (dicloxacillin flucloxacillin), methicillin, nafcillin, oxacillin and temocillin are narrow-spectrum antibacterials
Aminoglycosides which are broad-spectrum antibacterials include kanamycin A, amikacin, tobramycin, dibekacin, gentamicin, sisomicin, netilmicin, neomycins B, C and neomycin E (paromomycin) [7]	Cephalosporins (first generation and second generation) antibacterials are relatively narrow spectrum
Cephalosporins (third, fourth, and fifth generations) are relatively extended to the broad spectrum of activity Carbepenems (e.g. imipenems) show a broad pattern of activity [8]	Vancomycin, clindamycin, isoniazid, rifampin, ethambutol, pyrazinamide, bacitracin, polymixins, sulfonamides, glycopeptide and nitroimidazoles are counted in this group
Macrolides such as erythromycin, roxithromycin, clarithromycin, azithromycin, and dirithromycin are considered in this category [9]	
Tetracycline, chlortetracycline, oxytetracycline, demeclocycline, lymecycline, meclocycline, methacycline, minocycline, and tigecycline are considered as broad-spectrum antibacterials	
Chloramphenicol	
Ticarcillin, a carboxypenicillin, also has a broad spectrum of activity	
Rifamycins also exhibited broad coverage [10]	

Table 2. List of broad- and narrow-spectrum antibacterials.

been classified into two groups: group A (β-lactams) and group B (aminoglycosides). However, in a more elaborated way, the antibacterials can be classified into β-lactams, β-lactam/β-lactamase inhibitor combinations, aminglycoside, macrolides, quinolones, and flouroquinolones.

2.4.1. *β-Lactams*

Beta-lactams are a popular class of drugs, having a four-membered lactam ring (**Figure 1**), known as β-lactam ring; however, they vary by side chain attached or additional cycles. Penicillin derivatives, cephalosporins, monobactams, and carbepenems, e.g. imipenems, all belong to this class.

Usually, alterations were made to the basic penam and cephem structural units such that enhanced antimicrobial potential is achieved. Among such modified agents, some are clavulanate, latamoxef, loracarbef, etc. On the cephalosporins unit, most changes have been made at positions 7 and 3. Cephalothin, cephaloridine, and cephazolin are among some of the modified cephalosporins, which have shown good activity against Gram positive with the exception of enterococci- and methicillin-resistant staphylococci. Some other examples include preparation of microbiologically active oxacephems and carbacephems (**Figure 2**) by modification of the cephalosporin nucleus [11].

The aminopenicillins are also included in this class, which are structural analogues of ampicillin, which is a 2-amino derivative of benzylpenicillin [12].

2.4.2. *Aminglycoside*

In compounds of this group, two aminosugars joined by glycosidic bond to an aminocyclitol. Commonly used aminoglycosides are streptomycin, gentamicin, sisiomicin, netilmicin, kanamycin,

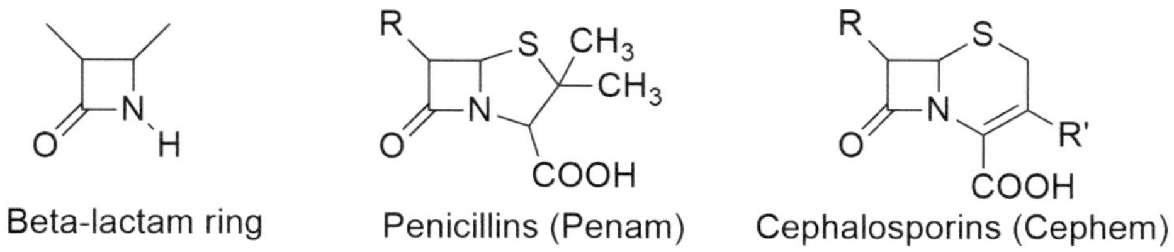

Figure 1. Basic structure of the β-lactam ring, penicillins (Penam skeleton) and cephalosporins (Cephem skeleton). R in Penam and Cephem nucleus represents the side chain that could be different for different penicillins and cephalosporins, while R′ denotes another side chain in the Cephem nucleus.

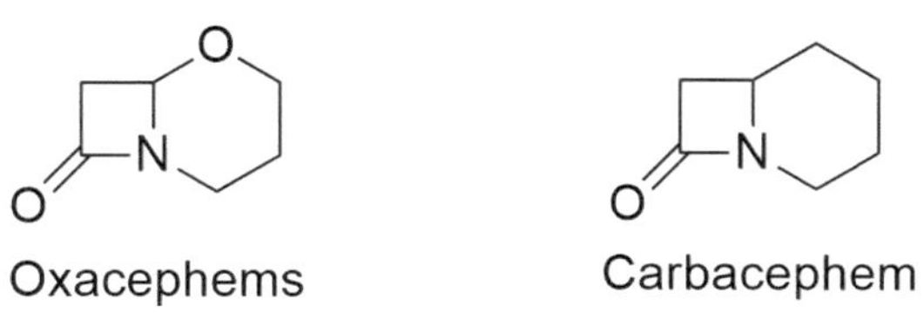

Figure 2. Cephalosporin-modified structure of oxacephems and carbacephems.

amikacin, neomycin, tobramycin, toframycin, spectinolycin, and paromonucin. The structure of some of these is presented in **Figure 3**.

Changes in original structural units of aminoglycosides can be made either synthetically or enzymatically. Structural properties such as the number and location of various functional groups on a modified compound compared to their parent compounds usually exhibit great effect on the biological activities of these drugs. The literature [13] has shown that the number and location of amino groups on the hexoses and the site of attachment of the other rings to deoxystreptamine have a considerable effect on preventing inhibition of protein synthesis or, in other words, their biological activities. For example, among kanamycin A, B, and C, kanamycin B is a highly effective antibiotic than either kanamycin A or C. It is inferred that the presence of a diamino hexose results in a compound that has better efficiency for inhibition of protein synthesis than the one holding only one amino group.

2.4.3. *Macrolides*

Macrolides belong to the polyketide class of natural products. Structurally, macrolides are antibiotics that consist of a macrocyclic lactone ring, usually 14-, 15-, or 16-member to which one or more deoxy sugars, usually cladinose and desosamine, may be attached. Some well-known examples of macrolides are erythromycin and roxithromycin etc.

So far, the relationship of structural activity of various macrolides has been studied. Studies revealed that some existing 14-, 15-, and 16-member macrolide antibiotics were modified toward interesting targets. For example, specific substitution on the C-9, C-11, C-12, or C-6 sites in the macrolactone ring results in better in vitro activity against mycobacterium tuberculosis (**Figure 4**) [14].

2.4.4. *Quinolones and flouroquinolones*

Quinolones are quinine-derived structural units and have been proved to be potent synthetic antibacterial agents. The basic skeleton of the quinolone molecule is presented in **Figure 5**. The addition of flourine at position 6 is called flouroquinolone. In the bicyclic ring, the variation

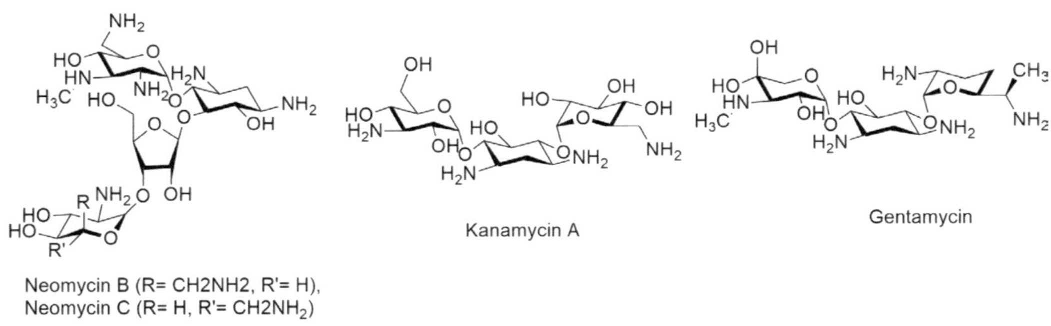

Figure 3. Structures of some well-known aminoglycosides antibacterials.

http://dx.doi.org/10.5772/intechopen.68695

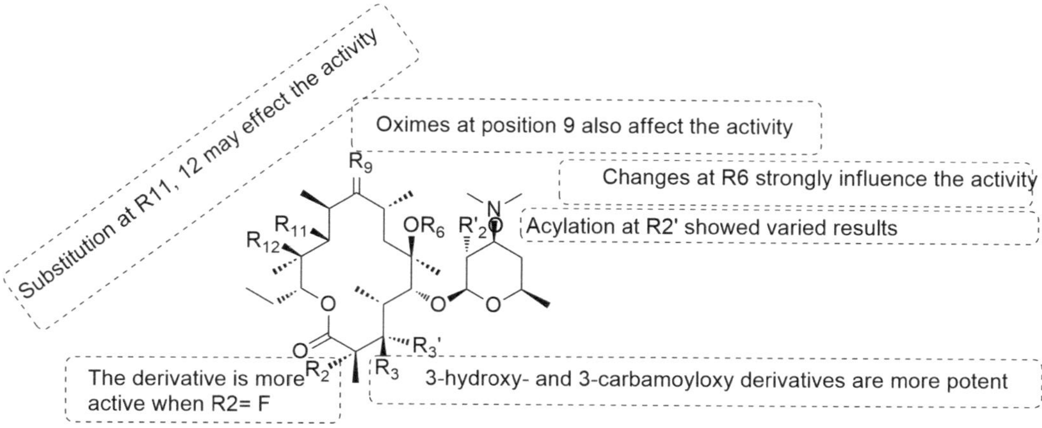

Figure 4. Substitution effect on activity of macrolides.

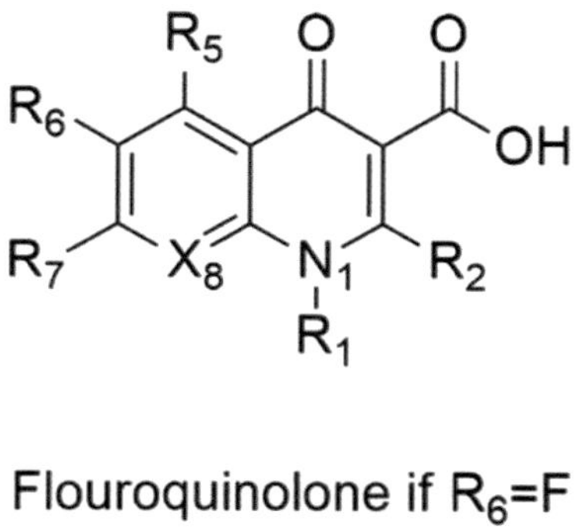

Figure 5. Basic structure of quinolone.

at positions 1-, 5-, 6-, 7-, and 8- exerts key effect on the therapeutic behaviour of these drugs. Usually, such structural alteration has led to enhanced coverage and potency of antibacterial activity and pharmacokinetics, e.g. improved anti-Gram-positive activity of moxifloxacin and garenoxacin. However, some of these modifications are associated with definite adverse effects [15]. Some well-known examples of quinolone include nalidixic acid (first generation), ciprofloxacin (second generation), levofloxacin (third generation), and trovafloxacin (fourth generation).

2.4.5. Streptogramin antibiotics

Streptogramin antibiotics are a unique class of antibacterials consisting of two groups of structurally unrelated molecules: group A streptogramins (polyunsaturated macrolactones) and group B streptogramins (cyclic hexadepsipeptides) [16]. Dalfopristin and quinopristin are representative example of the streptogramin A and streptogramin B groups, respectively. Alteration of the group B structural units has been mainly achieved on the 3-hydroxypicolinoyl, the 4-dimethylaminophenylalanine, and the 4-oxo pipecolinic residues. Modifications on this third part result in water-soluble derivative quinupristin. Water-soluble group A derivatives were obtained by some synthetic steps, e.g. dalfopristin, which is a sulfone derivative that can be obtained by Michael addition of aminothiols to the dehydroproline ring of pristinamycin IIA, followed by oxidation [17]. The group A molecules impede with the expansion of the polypeptide chain by avoiding the binding of aminoacetyl-tRNA to the ribosome and the creation of peptide bonds, while the group B building blocks encourage the disconnection of the peptidyl-tRNA and can interfere with the removal of the completed polypeptide by blocking its access to the channel through which it usually leaves the ribosome.

2.4.6. Sulphonamides

Sulphonamides are one of the important class of synthetic organic compounds with great medicinal importance having a sulphonamide functional group (R_1-SO_2-NR_2R_3) in their structures. Some compounds belonging to this group also show antibacterial properties such as sulfadiazine. The original antibacterial sulphonamides are synthetic antimicrobial agents that contain the sulphonamide group. Some others are sulfonylureas and thiazide diuretics which proved to be newer drug groups based on the antibacterial sulphonamides (**Figure 6**).

2.4.7. Tetracyclines

Tetracyclines are four rings hydrocarbon containing compounds, which can be defined also as "a subclass of polyketides having an octahydrotetracene-2-carboxamide skeleton." These antimicrobial agents were originally derived from *Streptomyces* bacteria, but the newer derivatives are semi-synthetic. Some promising examples of this group are oxytetracycline and doxycycline.

2.4.8. Nitroimidazoles

Nitroimidazoles are a group of compounds that contain a basic imidazole ring. The most commonly used example is metronidazole (**Figure 7**). Nitroimidazoles vary by the location of the nitro functional group. Most of the drugs of this class have their nitro group at position 6, such as metronidazole, and/or at position 2, such as benznidazole.

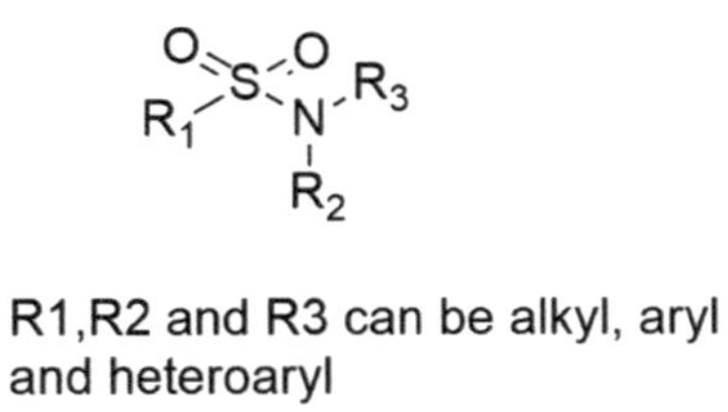

Figure 6. Basic structural unit of sulphonamide.

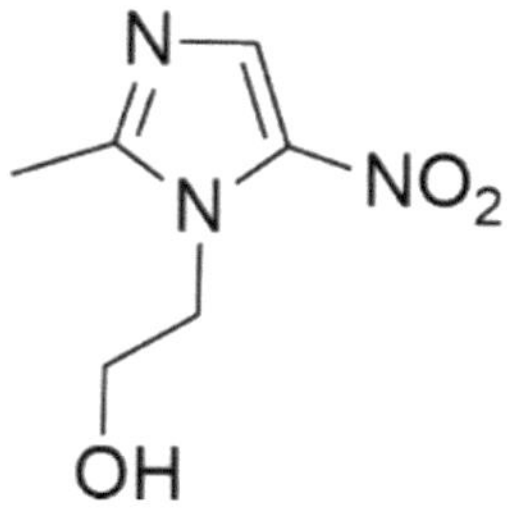

Figure 7. Structure of metronidazole.

2.5. Function-based classification of antibacterial drugs

Function means how a drug works or what is its mode of action. This is one of the most important factors related to each antibacterial. The major processes or functions, which are responsible for bacterial growth, are cell wall synthesis, cell membrane function, protein synthesis, nucleic acid synthesis, and so on. All such processes are targets for antibiotics; therefore, antibacterials, which interfere or disturb these processes in different ways, can be subdivided into four groups: such as cell wall synthesis inhibitors, inhibitors of membrane function, inhibitors of protein synthesis, and inhibitors of nucleic acid synthesis. All these groups are discussed briefly hereafter.

2.5.1. Cell wall synthesis inhibitors

Structurally, the bacterial cell wall is different from that of all other organisms by the presence of polysaccharide backbone, called peptidoglycan, which is composed of alternating *N*-acetylmuramic acid and *N*-acetylglucosamine residues in equal amounts and most of eubacteria have peptidoglycan-based cell walls except the mammalian cell. Like all other organisms, the bacterial cell wall offers structural completion to the cell; therefore, the most important process for avoiding bacterial growth is to stop cell wall synthesis by inhibiting the peptidoglycan layer of bacterial cell walls. The agents used to work against this function are called cell wall synthesis inhibitors and the cell wall of new bacteria growing in the presence of these agents is deprived of peptidoglycan.

β-Lactam drugs, including penicillin derivatives, cephalosporins, monobactams, and carbapenems, are the major antibiotics that inhibit bacterial cell wall synthesis. To understand the inhibition process, one must be aware of the fact that the last step in the synthesis of peptidoglycan is eased by penicillin-binding proteins; therefore, this initially occurs in the binding of drug to cell receptors, i.e. penicillin-binding proteins. Thus, β-lactam drugs work as a false molecule for D-alanyl-D-alanyl transpeptidases, which result in inhibition of transpeptidation reaction and peptidoglycan synthesis. Thereafter, autolytic enzyme inhibitors get inactivated, which activates the lytic enzyme, thereby resulting in division of bacteria provided that the environment is isotonic [18]. Some other antibiotics such as bacitracin, teicoplanin, vancomycin, ristocetin, and novobiocin must be subjected at early stages, which impede early phases of the peptidoglycan synthesis.

Gram-positive and Gram-negative bacteria vary in the susceptibility to the β-lactam drugs because of the structural differences in their cell wall, i.e. Gram-negative bacteria usually have less susceptibility because these antibiotics fail to reach the cell wall as they are blocked by the outer membrane of the Gram-negative bacteria. Factors such as the amount of peptidoglycan, receptors, and lipids availability, nature of crosslinking, autolytic enzymes activity greatly influence the activity, permeation, and incorporation of the drugs.

Considering the resistance phenomenon, all β-lactam antibacterials can only be inactivated by bacterial produced enzymes called β-lactamases (e.g. penicillinases, cephalosporinases, cephamycinases, carbapenemases, and so on).

2.5.2. *Inhibitors of membrane function*

The cytoplasmic membrane, which covers the cytoplasm, serves as a selective barrier and controls the internal composition of the cell. Whenever these functional roles of the cytoplasmic membrane get disturbed, macromolecules and ions will outflow, which will result in cell destruction or death. Selectivity of the agents is necessary to carry out this chemotherapy as the agents are aimed to target the bacterial cell membrane. Polymyxins are active antibacterial agents, which are cyclic peptides, having a long hydrophobic tail. Polymyxins are found in the form of A, B, C, D, E, where B and E can be used therapeutically. Polymysins show their specificity for polysaccharide molecules, which are present in the outer membrane of many Gram-negative bacteria; therefore, polymyxins are considered to be selectively toxic for Gram-negative bacteria. Mechanistically, after association with the lipopolysaccharide substrate in the outer membrane of Gram-negative bacteria, polymyxins change the membrane structure so that its permeability increases, which results in disruption of the osmotic balance. Additionally, changes like discharge of the molecules from interior of the cell, inhibition of respiration, and increased water uptake lead to the cell death. Since Gram-positive bacteria have a too thick cell wall, which denies the access of these molecules to the Gram-positive bacterial cell membrane, polymyxins have less or even no effect on Gram-positives [19].

2.5.3. *Protein synthesis inhibitors*

Protein synthesis is one of the most important functions in the bacterial cell and humans as well. Therefore, to cure infectious disease caused by pathogenic bacteria, it is the most important target for the drugs, which are called protein synthesis inhibitor antibiotics. Since both human and bacterial cells synthesize proteins, due to the slow synthesis of human proteins, it has remained a comfortable task for the development of the selective antibiotics. Only the side effects from toxicity and resistance phenomenon are taken seriously during antibiotic development.

Mechanistically, protein synthesis inhibitors act to disturb any stage of the protein synthesis such as initiation and elongation stages (aminoacyl tRNA entry, proofreading, peptidyl transfer, ribosomal translocation and termination). **Table 3** shows representative antibiotics, their sites and pathways, etc. [20].

2.5.4. *Inhibition of nucleic acid synthesis*

One of the most important targets for antibiotic to cure infectious diseases is nucleic acid synthesis, and the antibiotics used are called nucleic acid synthesis inhibitors. A sound

Drug type	Binding site function and pathway disturbed
Aminoglycosides: Examples include gentamicin, tobramycin, streptomycin, and kanamycin	Aminoglycosides bind to the 30S ribosomal subunit which alter the ribosomal structure. This affects all normal steps of protein synthesis, such as initiation step of translation, blocking of elongation of peptide bond formation, discharge of incomplete, and toxic proteins. These disturbances ultimately stop protein synthesis and destroy the cytoplasmic membrane.
Macrolides	These are protein synthesis inhibitors, which bind to the 50S ribosomal subunits, impeding peptidyl transfer
Tetracyclines and glycyclines (tigecycline)	These inhibitors bind to the 30S ribosomal subunit. Protein translation (through inhibition of aminoacyl tRNA binding to ribosome) gets disturbed by these inhibitors
Strptogramines: Examples include pristinamycin, dalfopristin, and quinupristin	Their binding site is the 50S ribosomal subunit. They interfere in protein translation through prevention of initiation, elongation, and translocation stages and free tRNA depletion
Phenicols: For example, chloroamphenicol	These antibiotics, e.g. chloroamphenicols, bind to the 50S ribosomal subunit and inhibit protein synthesis by blocking the peptidyl transfer phase of elongation on the 50S ribosomal subunit in bacteria
Oxazolidinone: The most common example is linezolid	They bind to the 50S ribosomal subunit, which are thought to act at the initiation stage [21]
Ketolides: This is a novel class of protein synthesis inhibitors, which exhibit excellent activity against resistant organisms.	
Protein synthesis inhibitors with unknown pathway include retapamulin, mupirocin, and fusidic acid.	

Table 3. Example of drugs, their binding sites and pathways which get affected.

difference in the enzymes that carry out DNA and RNA synthesis between eukaryotic and prokaryotic cells helps to achieve selective toxicity, which favours development of the antibiotic. The antibacterials of this class can be subdivided into DNA inhibitors and RNA inhibitors. RNA inhibitors interfere with the bacterial transcription process in which messenger RNA transcripts of genetic material are produced for later transformation into proteins. RNA inhibitors such as rifampin, a well-known example of the rifamycins family, bind to DNA-dependent RNA polymerase, thereby creating a wall that inhibits elongation of RNA. Such a situation prevents gene transcription which affects the normal function of bacteria that results in cell death. Like all other biological polymerization processes, DNA synthesis is also achieved by initiation, elongation, and termination stages; therefore, antibacterial drugs target any one of these processes to inhibit DNA synthesis. Quinolones, including nalidixic acid and ciprofloxacin, work as DNA inhibitors. DNA gyrase (a topoisomerase) is accountable for cutting one of the chromosomal DNA parts at the beginning of the supercoiling. The scratch is made provisionally and later on linked back together. Quinolones bind to DNA gyrase, inhibiting their function, which results in inhibition of the DNA replication that ultimately results in cell damage. There are some other antibacterial drugs, which act upon anaerobic bacteria by creating metabolites that are bind into DNA strands, which then are more likely to rupture. Examples of such drugs include nitrofurantoin and metronidazole.

3. Recent antimicrobial agents

Our discussion covered almost all the old and some new antimicrobial agents. However, to make these agents easily understandable, **Table 4** lists some recent antibacterial agents with their structure, class, and so on [22].

FDA-approved antibacterial agent	**Structure**	**Category**	**Approval year/ trial phase**
Tigecycline	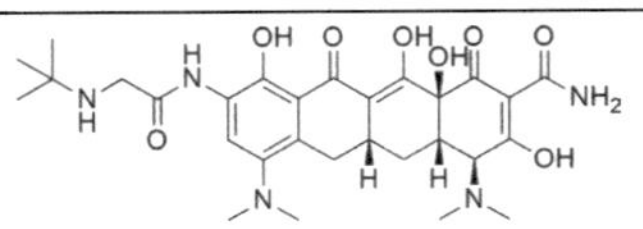	Glycylcycline	2005
Doripenem	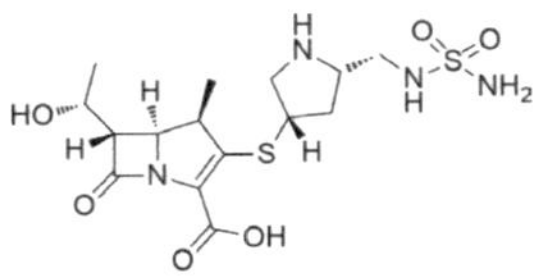	Carbapenems	2007
Retapamulin	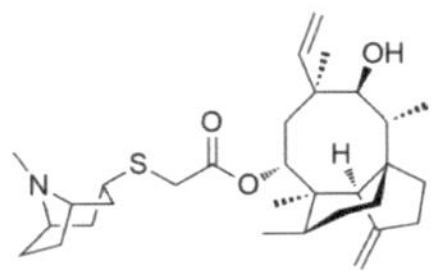	Pleuromutillin	2007
Telavancin	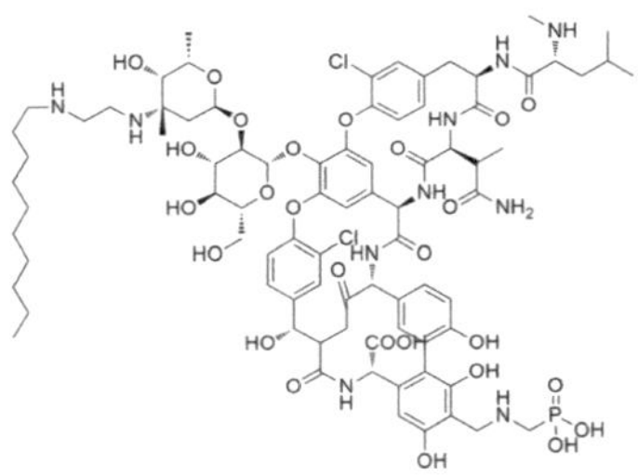	Glycopeptides	2009
Ceftaroline	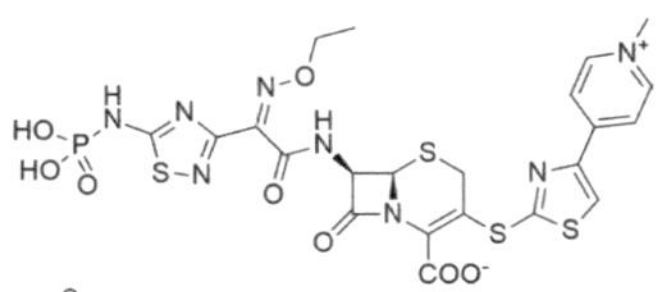	Cephalosporins	2010
Fidaxomicin	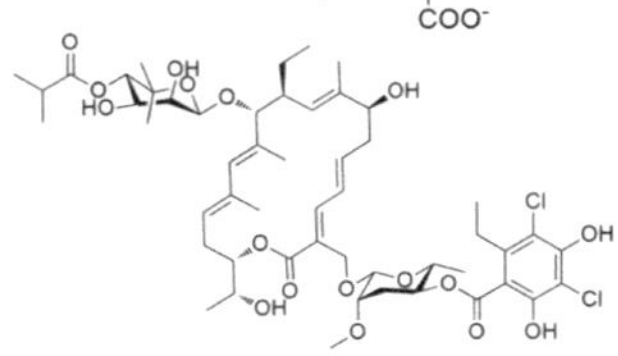	Macrocyclic	2011
FDA approval awaiting antibacterial agents	Structure	Category	Approval year/ trial phase
Ceftobiprole	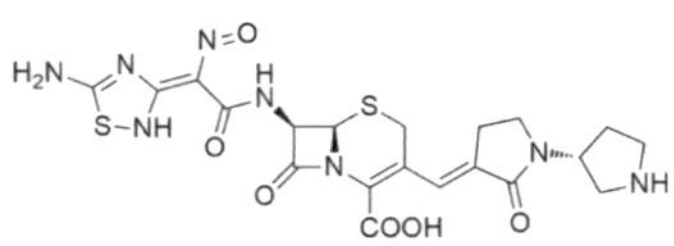	Cephalosporin	Awaited

http://dx.doi.org/10.5772/intechopen.68695

FDA-approved antibacterial agent	Structure	Category	Approval year/ trial phase
Iclaprim	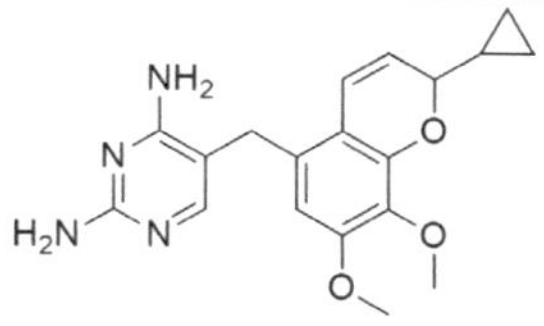	Dihydropholate reductase inhibitor	Awaited
Torizolid	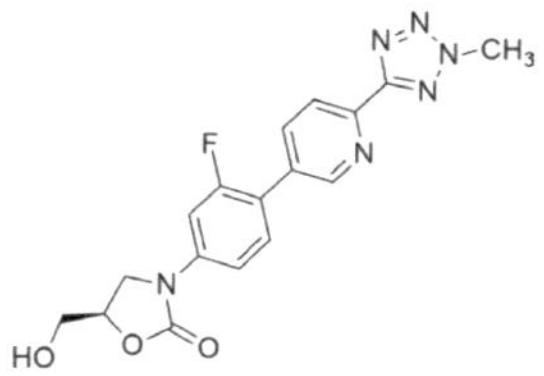	Oxazolidinones	Phase II
Radezolid	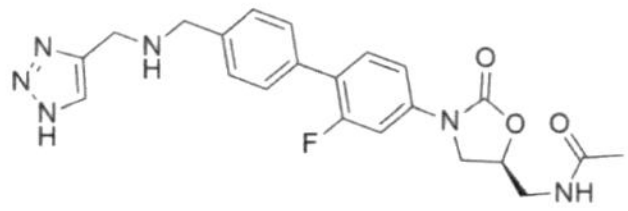	Oxazolidinones	Phase II
Cethromycin	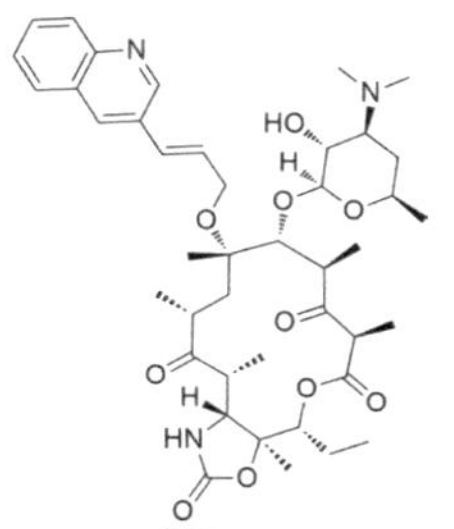	Ketolides	Phase III
Solithromycin	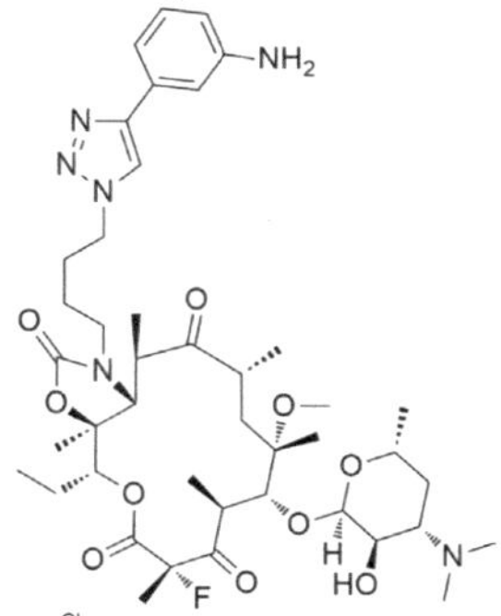	Ketolides	Phase II
Oritavancin	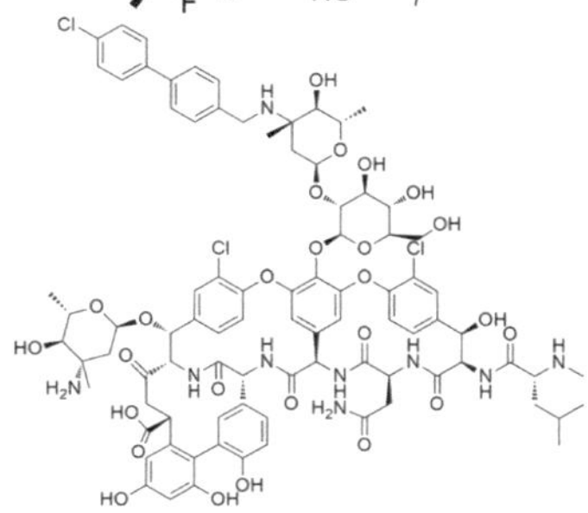	Glycopeptide	Phase III

FDA-approved antibacterial agent	Structure	Category	Approval year/ trial phase
Dalbavancin	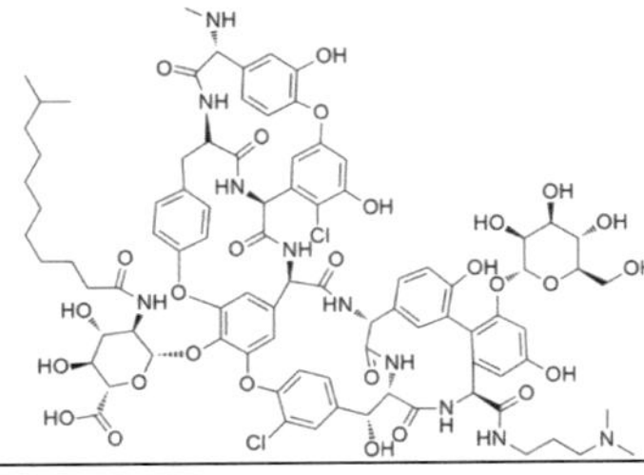	Glycopeptide	Phase III

Table 4. List of newer antibacterial agent.

4. Conclusion and prospectives

Unlike antibiotics classification, little efforts have been made to classify antibacterials (a subclass of antibiotic) separately. Therefore, we tried to classify antibacterial into five principal categories, each of which has its own importance. However, classifications based on chemical structure and function of these agents are considered to be more important as these groups describe a lot about their therapeutic nature, while the rest of the classification is less important, e.g. sometimes, classification based on the spectrum of activity distinguishes these agents in an ambiguous way as the spectrum sometime depends on their concentration used. The classification mentioned could be a better guide for future classification, i.e. the agents that are in developing stages or those that are going to develop can be adjusted in any suitable group mentioned in the text. Further, this categorization could be helpful in academic and in health care fields at present and in the future as well.

Author details

Hamid Ullah[1,2]* and Saqib Ali[2]

*Address all correspondence to: hamidullah9@gmail.com

1 Department of Chemistry, Faculty of Arts and Basic Sciences, Balochistan University of Information Technology, Engineering and Management Sciences (BUITEMS), Quetta, Pakistan

2 Department of Chemistry, Mohi-ud-Din Islamic University, AJ&K, Pakistan

References

[1] Adzitey F. Antibiotic classes and antibiotic susceptibility of bacterial isolates from selected poultry. World's Veterinary Journal. 2015;**5**:36-41. pii: S232245681500008-5

[2] Aminov RI. A brief history of the antibiotic era: Lessons learned and challenges for the future. Frontiers in Microbiology. 2010;**1**:1-5. DOI: 10.3389/fmicb.2010.00134

[3] Oloke JK Activity pattern of natural and synthetic antibacterial agents among hospital isolates. Microbios. 2000;**102**:175-181

[4] Acar J. Broad- and narrow-spectrum antibiotics: An unhelpful categorization. Clinical Microbiology and Infection. 1997;**3**:395-396. DOI: 10.1111/j.1469-0691.1997.tb00274.x

[5] Carbon C, Isturiz R. Narrow versus broad spectrum antibacterials: Factors in the selection of pneumococcal resistance to beta-lactams. Drugs. 2002;**62**:1289-1294. DOI: 10.2165/00003495-200262090-00001

[6] King DE, Malone R, Lilley SH. New classification and update on the quinolone antibiotics. American Family Physician. 2000;**61**:2741-2748

[7] Kotra LP, Haddad J, Mobashery S. Aminoglycosides: Perspectives on mechanisms of action and resistance and strategies to counter resistance. Antimicrobial Agents Chemotherapy. 2000;**44**:3249-3256. DOI: 10.1128/AAC.44.12.3249-3256.2000

[8] Zhanel GG, Wiebe R, Dilay L, Thomson K, Rubinstein E, Hoban DJ, Noreddin AM, Karlowsky JA, Hoban DJ, Noreddin AM, Karlowsky JA. Comparative review of the carbapenems. Drugs. 2007;**67**:1027-1052. DOI: 10.2165/00003495-200767070-0000

[9] Hof H. Macrolides, a group of antibiotics with a broad spectrum of activity. Immunität und Infektion.1994;**22**:66-71

[10] Floss HG, Yu TW. Rifamycins mode of action, resistance, and biosynthesis. Chemical Review. 2005;**105**:621-632. DOI: 10.1021/cr030112j

[11] Hamilton MJMT. β-Lactams: Variations on a chemical theme, with some surprising biological results. Journal of Antimicrobial Chemotherapy. 1999;**44**:729-734. DOI: 10.1093/jac/44.6.729

[12] Cunha BA. Aminopenicillins in urology. Urology. 1992;**40**:186-190. DOI: 10.1016/0090-4295(92)90525-2

[13] Benveniste R, Davies J. Structure-activity relationships among the aminoglycoside antibiotics: Role of hydroxyl and amino groups. Antimicrobial Agents Chemotherapy. 1973;**4**:402-409. DOI: 10.1128/AAC.4.4.402

[14] Zhu ZJ, Krasnykh O, Pan D, Petukhova V, Yu G, Liu Y, Liu H, Hong S, Wang Y, Wan B, Liang W, Franzblau SG. Structure activity relationships of macrolides against *Mycobacterium* tuberculosis. Tuberculosis. 2008;**88**:49-63. DOI: 10.1016/S1472-9792(08)70036-2

[15] Emami S, Shafiee A, Foroumadi A. Quinolones: Recent structural and clinical developments. Iranian Journal of Pharmaceutical Research. 2005;**4**:123-136

[16] Mast Y, Wohlleben W. Streptogramins – Two are better than one. International Journal of Medical Microbiology. 2014;**304**:44-50. DOI: 10.1016/j.ijmm.2013.08.008

[17] Barrière JC, Berthaud N, Beyer D, Dutka-Malen S, Paris JM, Desnottes JF. Recent developments in streptogramin research. Current Pharmaceutical Design. 1998;**4**:155-180

[18] Bugg TDH, Braddic D, Dowson CG, Roper DI. Bacterial cell wall assembly: Still an attractive antibacterial target. Trends in Biotechnology. 2011;**29**:167-173. 10.1016/j.tibtech.2010.12.006

[19] Newton BA. Mechanisms of antibiotic action. Annual Review of Microbiology.1965;**19**: 209-240. DOI: 10.1146/annurev.mi.19.100165.001233

[20] Kohanski MA, Dwyer DJ, Collins JJ. How antibiotics kill bacteria: From targets to networks. Nature Reviews Microbiology. 2010;**8**:423-435. DOI: 10.1038/nrmicro2333

[21] Swaney SM, Aoki H, Ganoza MC, Shinabarger DL. The oxazolidinone linezolid inhibits initiation of protein synthesis in bacteria. Antimicrobial Agents Chemotherapy. 1998;**42**: 3251-3255

[22] Rai J, Randhawa GK, Kaur M. Recent advances in antibacterial drugs. International Journal of Applied and Basic Medical Research. 2013;**3**:3-10. DOI: 10.4103/2229-516X.112229

Chapter 2

Novel Antimicrobial Agents and Processes for Textile Applications

Monica Periolatto, Franco Ferrero, Claudia Vineis,
Alessio Varesano and Giuseppe Gozzelino

Additional information is available at the end of the chapter

http://dx.doi.org/10.5772/intechopen.68423

Abstract

The use of antimicrobial compounds in textiles has grown dramatically over the last decades. The potential application field is wide. It ranges from industrial textiles exposed to weather such as awnings, screens and tents; upholstery used in large public areas such as hospitals, hotels and stations; fabrics for transports; protective clothing and personal protective equipment; bed sheets and blankets; textiles left wet between processing steps; intimate apparel, underwear, socks and sportswear. Another large field of application is in filtration and disinfection of air and water for white rooms, hospitals and operating theatres, food and pharmaceutical industries, water depuration, drinkable water supplying and air-conditioning systems. The present chapter is a review of recent research works related to antimicrobial finishes for textile materials. Several examples of antimicrobial treatments (e.g. traditional pad-dry-cure technique, exhaustion bath, encapsulation, electrospinning, cross-linking, etc.) were reported. The antimicrobial agents were divided by their origin from synthesis or from natural sources. Quaternary ammonium compounds (QACs), Triclosan, metals (including metal oxides and salts), polyhexamethylene biguanide (PHMB), N-halamines and conjugated polymers (i.e. polypyrrole) were listed as synthetic biocides in textile applications. Extracts from plants (e.g. aromatic compounds, essential oils and dyes), antimicrobial peptides (AMPs) and chitosan were considered among natural-based biocides.

Keywords: textiles, cotton, chitosan, polypyrrole, antimicrobial, photo-grafting

1. Introduction

Fibres, both natural and man-made, have been widely used since the ancient past in the manufacture of other materials. World fibre consumption has strongly increased over the years, reaching a total demand of 94.9 million tons in 2015. In detail, 66.8 million tons were man-made fibres, in addition to natural fibres with a demand of 28.1 million tons [1].

Nowadays, besides the traditional clothing products, textiles find important applications also in home furnishing, food packaging, as fibre reinforcements for polymers, optical fibres, thermal and mechanical protection, sport equipment, fibrous materials for a large array of applications in medicine and hygiene such as medical devices, health care and hygienic coatings, air filters and water purification systems.

An important example of these functional fabrics, recently attracting the interest of the research, is antimicrobial fabrics. Due to the morphology of fibres, in particular those of natural origin, textiles are prone to microorganisms' growth on their surface, due to the large surface area and moisture affinity. Bacteria and fungi can be found everywhere, so the contact with textiles is extremely probable. Depending on moisture, nutrients, temperature and pH, their growth can be very fast: some bacteria can double every 20 min [2].

The undesirable effects caused by microorganisms' growth act both on the textile itself and on the user. Unpleasant odour, reduction of mechanical strength, stains and discolouration are all effects of the biodeterioration of textiles, affecting almost all the types of fibres. Natural fibres are generally more susceptible to biodeterioration than the man-made fibres, because their porous hydrophilic structure retains water, oxygen and nutrients, providing perfect environments for bacterial growth. Finishing agents can also promote microbial growth [3]. Even mild surface growth can make a fabric unattractive by the appearance of unwanted pigmentation; heavy infestation which results in rotting and breakdown of the fibres may cause the fabric deterioration, in fact microorganisms can accelerate the hydrolysis of cellulose. Man-made fibres derived from cellulose are susceptible to microbial deterioration. Viscose is readily attacked by mildews and bacteria while acetate and triacetate are more resistant, but discolouration can occur if the fabrics are incorrectly stored. Synthetic fibres show strong resistance to attack by microorganisms, due to the hydrophobic nature of the polymers, but the presence of contaminants can cause some bacterial attack [3].

Most of the microorganisms involved in textile contaminations can cause pathogenic effects. Many species, such as *Escherichia coli, Klebsiella pneumonia, Pseudomonas aeruginosa, Staphylococcus aureus and Acinetobacter baumannii,* can cause infections to human beings due to user contaminations. It is a great concern for textiles used mainly in hospitals, as medical devices or for health and hygienic care, and in crowded places, but also in sport and underwear clothing, water purification systems, animal feed and food industry. Therefore, the demand for antimicrobial textiles is gaining interest, showing a strong increase over the last few years: the global market for antimicrobial agents is expected to increase by about 12% each year between 2013 and 2018 [4].

Different methods were investigated to confer antimicrobial activity to textiles; they can be classified into the inclusion of antimicrobial leaching compounds in the polymeric fibres, the surface modification of the fibres by grafting reactions or by physical methods.

Besides the efficiency towards a broad spectrum of microorganisms, any antimicrobial treatment should consider other challenges. First, it has to be non-toxic to the end user of the textile, namely cytotoxicity, allergy, irritation or sensitization must be avoided. The treatment must have a good fastness to use, mainly to repeated laundering, dry cleaning and ironing, and should not compromise the quality, hand or appearance of the textile. The application method should be simple, easily implementable in the finishing process and environmentally friendly, avoiding side effects for the manufacturers. Finally, the antimicrobial agent should not kill the resident flora of non-pathogenic bacteria on the skin of the wearer. Thus, the study on new and efficient antimicrobial treatments for textiles, considering both the antimicrobial agent and the application method, is a relevant topic of the research.

The aim of this chapter is to provide an overview of recently developed antimicrobial treatments to produce antimicrobial textiles. Afterwards, the discussion will be focused and detailed on chitosan and polypyrrole (PPy), two promising antimicrobial agents deeply investigated by the authors for textile applications.

2. Antimicrobial treatments for textiles

Depending on the fibre type, that is morphology, composition and surface texture, and on the applied antimicrobial agent, different chemical or physical approaches are possible and under development to confer antimicrobial activity to textiles.

In the case of synthetic fibres, a specific antimicrobial agent can be directly incorporated into the polymeric matrix [5].

The application of the antimicrobial agent on the fibre's surface, during the finishing stage, is a viable method both for synthetic and natural fibres; it can be carried out by the traditional pad-dry-cure technique or exhaustion bath.

The recent growing interest on nanotechnology concerns also the textile field; in fact nanoscale particles can be prepared from natural or synthetic compounds with antibacterial activity and applied to textiles, for example, by foulard. The advantage is the lower add-on enough to confer the desired property due to the high surface area of the nanoparticles. Moreover, coupling the process with a final cross-linking, a good fastness of the finishing can be obtained [6]. Electrospinning to produce intrinsically antimicrobial nanofibres is another interesting application of nanotechnology; in this case, the nanofibres can be coupled with other natural or synthetic fibres to produce antimicrobial yarns [7].

Microencapsulation is a process by which droplets of liquid or particles of solid are covered with a continuous film of polymeric material [8]. This technology is one of the most promising techniques to confer functional properties to textiles: the capsules are applied to fibres as dispersion with a binder using padding, spraying, impregnation, exhaust or screen-printing techniques. It is more advantageous than the conventional processes in terms of economy, energy saving, eco-friendliness and controlled release of substances, but it can affect the handle of the textile [9].

An effective way to embed the antimicrobial agent in the fibre is cross-linking. Cross-linking happens when a cross-linker makes intermolecular covalent bridges between the polymer chains and the antibacterial molecule. Cross-linkers include glutaraldehyde, genipin, glyoxal, dextran sulphate, 1,1,3,3-tetramethoxypropane, oxidized cyclodextrins, ethylene glycol diglyceryl ether, ethylene glycol diglycidyl ether (EGDE) and diisocyanate [10, 11].

Cross-linking can occur by chemical [12], radiation [13] or physical method [14]. In radiation, cross-linking, heat or a catalyst are not needed, thus no additional toxic chemical is introduced into the system and the substrate is preserved by a thermal degradation. The physical method is based on ionic interactions between polymer chains, so it is not as durable as the chemical or radiation ones.

Finally, altering the surface properties of fibres is also an interesting way to ensure a strong adhesion of finishing agents to textiles. Surface modification methods, such as oxygen plasma treatment, ultrasound technology, UV radiation, surface bridging and enzyme treatment, have been recently investigated, with the aim to impart durable antimicrobial finishes to fabrics using mainly natural products [15].

Depending on the application method, the antimicrobial textile can act by contact or by diffusion. In the first case, the antimicrobial agent is placed on the surface of the substrate and no leaching occurs; it will act only in case of direct contact between the microorganism and the fibres. In the second case, the agent will migrate from the textile to the external environment, to attack the microorganisms. It means that the antimicrobial activity of the textile can decrease with time, and that the impact of the antimicrobial agent on the environment has to be considered.

3. Antimicrobial agents of synthetic origin

According to its action against the microorganisms' cell, an antimicrobial agent can be classified as biostatic or biocidal. The first ones can just inhibit the cell growth, whereas biocidal agents can kill the microorganisms. Most of the antimicrobial agents used in commercial textile finishing are biocides, acting by damage or inhibition of cell wall synthesis, inhibition of cell membrane function, of protein synthesis, of DNA and RNA synthesis or of other metabolic processes.

Quaternary ammonium compounds (QACs) are cationic agents carrying a positive charge at the N atom in solution ($R_4N^+X^-$); they are usually attached to an anionic fibre surface (polyester, cotton, nylon and wool) by ionic interaction. The molecule is a linear alkyl ammonium chain composed of a hydrophobic alkyl chain (C12–C18) and a hydrophilic counterpart.

The antimicrobial action, depending on alkyl chain length, presence of perfluorinated groups and cationic ammonium group, is due to the interaction of positive charges on the surface and cell membrane negative charges, with the consequent loss of membrane permeability and cell leakage. It causes the damage of cell membranes, the denaturation of proteins and the inhibition of DNA production [16].

QACs are effective against Gram-positive and Gram-negative bacteria, fungi and certain types of viruses; for this reason, these are widely used in industrial applications [17]. The disadvantage is the poor fastness of the treatment due to the fast leaching from the textile for the lack of chemical or physical bonding [18].

Commercial products based on QAC are BIOGUARD® (AEGIS Microbe Shield, New Zealand), Sanigard KC® (LN Chemical Industries, Switzerland) and Sanitized® (SANITIZED, Switzerland) [19].

Triclosan is a 2,4,4′-trichloro-2′hydroxydiphenyl ether ($C_{12}H_7Cl_3O_2$), a synthetic chlorinated bisphenol not ionized in solutions, improving its durability to laundering. It can act against Gram-negative and Gram-positive bacteria and against some fungi and viruses [5, 20] by blocking lipid biosynthesis affecting the integrity of cell membranes [21].

Triclosan has become, in last decades, the most efficient and widely used bisphenol in many application fields. On textiles, it is mainly used in association with polyester, nylon, polypropylene, cellulose acetate and acrylic fibres [22].

This recent widespread use of Triclosan had the drawback to generate bacterial resistance. Moreover, the reported photochemical conversion of Triclosan to 2,8-dichlorodibenzo-p-dioxin in aqueous solutions is another great concern, due to its toxicity [23].

Commercial products based on Triclosan, either as an isolated agent for a finishing or incorporated in fibres, are Microban® (Cannock, United Kingdom), Irgaguard® 1000 (Ludwigshafen, Germany), BiofresH™ (Salem, MA, USA) and Silfresh® (Magenta, Italy).

Metals, oxide or salt compounds, based on silver, copper, zinc or cobalt, have a strong biocidal effect due to the metal reduction potential, metal donor atom selectivity and speciation. These compounds can bind to O, N or S donor ligands present in the microorganism cell, inducing an oxidative stress, damaging cellular proteins, lipids and DNA.

Among all, silver particles were widely exploited due to the broad spectrum of action against bacteria like *P. aeruginosa, S. aureus, S. epidermidis, E. coli and K. pneumoniae*. In textile fields, they are mainly applied in the form of salts (79%) rather than metallic (13%) or ionic (8%) form [24]. Recently, the application in the form of nanoparticles, obtained by sol-gel, is gaining interest for silver, CuO, ZnO and TiO_2 [25]. It is due to the higher surface area with respect to larger particles, higher solubility and faster release of the metal ions, turning in a stronger antimicrobial effect. On ZnO, it was found that the antibacterial activity is inversely proportional to the nanoparticle size [26].

The scale-up of the process to commercial scale, unfortunately, was contained due to cost, environmental and technical challenges. A plasma, UV or acidic pre-treatment is often required on fabrics to improve the treatment durability, otherwise not so good [27].

Available commercial products are mainly based on silver, in isolated form, for fibre finishing or incorporation, or already in fibre or fabric form. Some examples are Ultra-Fresh® and Silpure®, SmartSilver®, MicroFresh® and SoleFresh®, Bioactive® and Silvadur™.

Polyhexamethylene biguanide (PHMB) ($(C_8H_{17}N_5)_n$) is a polycationic amine in which the cationic biguanide groups are interdispersed between hydrophobic hexamethylene groups. Electrostatic and hydrophobic interactions occur with microbial cell membranes, resulting in cell membrane disruption and lethal leakage of cytoplasmic materials. Its antibacterial activity increases with the level of polymerization [28]. Some PHMB-based textile products, such as Biozac ZS and Reputex®, have already appeared on the market as finishing products [29].

N-halamines are heterocyclic organic compounds, with one or two covalent bonds between nitrogen and a halogen, usually chlorine (N–Cl). N-halamines can be imide, amide or amine depending on the covalent bonds formed; the antimicrobial activity increases in the same order, while the stability decreases. N-halamines present a biocide action against a broad spectrum of bacteria, fungi and viruses, binding to the acceptor regions on microorganisms, precluding the cell enzymatic and metabolic processes and causing the consequent microorganism destruction [30]. Besides the low cost and wide range action, an advantage is the possibility to recharge their antimicrobial effect of the inactive substance by simply reacting them with Cl donor compounds [31]. As a disadvantage, textiles' treatment with N-halamine may result in a substantial amount of adsorbed Cl on the fibre surface. Those residues may produce an unpleasant odour or even discolour fabrics, which is a concerning disadvantage to the textile industry.

Conjugated polymers, such as **polypyrrole**, are generally applied in textile field as electrically conductive coating in order to produce electrically conductive textiles [32, 33]. PPy can be easily produced by chemical oxidative polymerization from water solutions of the monomer. Textile materials (e.g. fibres, yarns and fabrics) soaked in the polymerization bath are coated with an even and uniform layer of PPy by in situ chemical oxidative polymerization. During the oxidative polymerization, positive charges are introduced along the backbone chain of PPy. The charges are counter-balanced by counter-ions (also called dopants or doping agents), namely anions present in the polymerization bath. The anions in the polymerization bath are embedded in the polymer matrix improving the formation and stability of positive charges along the backbone chain of PPy, delocalized over several monomer units. PPy has been a subject of several works that evaluate its properties as biocidal agent. Excellent antimicrobial properties have been shown against both Gram-negative and Gram-positive bacteria. Such a bioactivity of PPy is likely due to the presence of positive charges, even if no leaching of biocidal substances has been proven on PPy-coated fabrics. The 'non-leaching' approach would avoid or limit the release of toxic biocide agents to the environment or to the skin of the wearers, in the case of garments.

PPy nanoparticles were synthesized by chemical polymerization using ammonium persulphate as oxidant following different methods in order to evaluate the influence on the morphology of resulting nanoparticles and bactericidal activity [34]. Five systems were synthesized: conventional PPy (without surfactants), highly soluble PPy (in SDS solution), PPy/Ag colloid (in PVA solution), branched PPy and branched PPy/Ag nanocomposite (in CTAB/

SDBA solution). Resulting polymer particles were investigated as active bactericidal materials against *E. coli, S. aureus and K. pneumoniae*. Inhibition zones, minimum inhibitory concentration (MIC) and time-kill assays were evaluated. The results indicate that the incorporation of silver nanoparticles improves the biocidal action of PPy and confirm that the size of PPy nanoparticles represents a relevant parameter for the bactericidal activity. In particular, it is possible to list the biocidal activity as follows: highly soluble PPy > branched PPy with silver > branched PPy > colloidal PPy > conventional PPy.

Antimicrobial activity of PPy on fabrics was first reported by Seshadri and Bhat [35] in 2005. In particular, they deposited PPy on cotton fabrics by in situ chemical oxidative polymerization at cold temperature (0–5°C). The fabrics were impregnated with monomer solutions and then the oxidant solution (ferric chloride) was added producing PPy-coated fabrics. $CuCl_2$ was used to treat samples of PPy-coated fabrics as an additional antimicrobial agent. The biocidal properties were tested by AATCC Test Method 147-1993 and ASTM E 2149-01 using *S. aureus, E. coli* and *Candida albicans*. The microbial reductions were 65, 59 and 73%, respectively. The addition of $CuCl_2$ to PPy increased the biocidal efficiency to 93, 98 and 100%, respectively.

In another paper [36], cotton fabrics were coated with PPy at room temperature using different oxidizing agents in order to assess their antimicrobial efficacy. The fabrics were soaked in a water solution of the oxidant. The monomer was added drop-wise to the stirred bath, and the reaction lasted for 4 h producing an even black layer of PPy on the fibres of the fabrics (**Figure 1**).

To obtain information about the influence of the oxidation agents on the biocidal activity, the synthesis of PPy was carried out using three different oxidants: ferric chloride, ferric sulphate and ammonium persulphate.

With ferric chloride and ferric sulphate, PPy is produced by the redox reaction between the monomer and ferric ions, reduced to ferrous ions. Using ammonium persulphate, the oxidative component is persulphate ion reduced to $SO_4^{=}$. Using ferric sulphate and ammonium persulphate, PPy embeds $SO_4^{=}$ ions as counter-ions, whereas the PPy produced with ferric

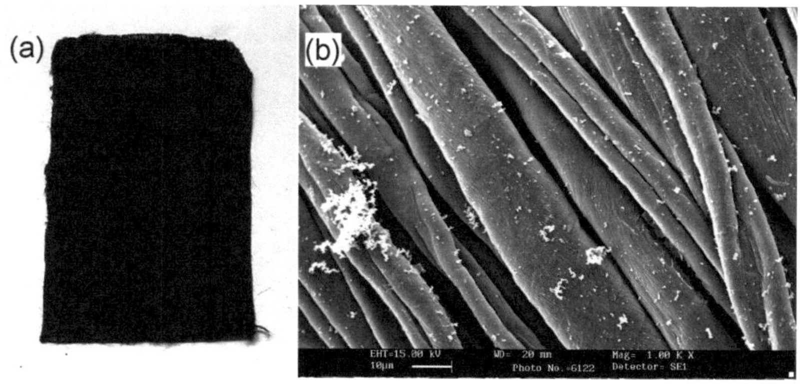

Figure 1. (a) Picture of the PPy-coated cotton fabric and (b) SEM image of cotton fibres coated by PPy.

chloride embeds Cl^-. Both ferric chloride and ferric sulphate give a high acidic pH to the polymerization bath due to the production of ferric complexes with OH^-. On the contrary, solutions of ammonium persulphate have a relatively low pH, due to the hydrolysis equilibrium of ammonium ions and water.

Antibacterial activity of PPy-coated fabrics was evaluated following the ISO 20645:2004 procedure using *E. coli* by placing the fabrics in contact with bacteria. No inhibition zone was observed after 24 h and the colonies grew around the fabric. The absence of colonies was observed under the fabrics in the contact zone. Therefore, there is an antibacterial activity on the fabric surface just by contact because PPy cannot diffuse being linked to the fabric. The absence of bacterial growth, even without inhibition zone, may be considered as a good antibacterial compound. Finally, the results pointed out that the antibacterial property is independent on either the oxidant used in the synthesis of PPy or the dopant embedded in the polymer matrix.

Cotton fabrics have also been coated with PPy using dicyclohexyl sulphosuccinate (DSS) [37]. DSS has two functions: (a) it is embedded into PPy as counter-ion, similarly to several other dopants with an $-SO_3^-$ group, (b) it greatly enhances the deposition process of PPy on the fabric by lowering the surface tension as a surfactant, and in turn it increases the evenness and weight uptake of PPy. In fact, the weight uptake of PPy has been 12% without DSS and 18% with DSS. The difference has been attributed primarily to the more efficient deposition due to the increased wetting of fibre surface caused by the surfactant action of DSS.

Antibacterial activity of the PPy-coated fabrics has been evaluated following the ASTM E 2149-01 procedure. Both the fabrics coated with PPy without DSS and PPy with DSS show 100% bacterial reduction, while untreated cotton fabric had practically no antibacterial activity. The stability of the biocidal action has been evaluated after different kinds of laundering. In particular, after dry-cleaning, fabrics coated with PPy without DSS and PPy with DSS show high bacterial reductions, 99 and 98%, respectively, whilst antibacterial efficiency decreases after launderings with non-ionic and anionic surfactants. In particular, the antibacterial activity of fabrics coated with PPy without DSS has been severely degraded by anionic laundering.

Moreover, the paper evaluated the biocidal mechanism of PPy by carrying out scanning electron microscopy (SEM) analysis of *E. coli* bacteria on cotton fibres and PPy-treated fibres. Bacteria on untreated cotton fibres had typical and regular bacterial shapes indicating that cells survived on the fibre surface. On the contrary, *E. coli* cells on PPy-coated fibres showed altered shapes probably due to the opening of their membrane and leakage of intracellular components from bacterial cells.

Recently, PPy has also been used for antimicrobial applications in combination with silver [38, 39]. In particular, Omastová et al. [40] prepared polypyrrole/silver composites by a single-step chemical oxidative polymerization using silver nitrate as an oxidant in water at room temperature. The reaction needed several days in order to yield more than 70%. The silver content in the PPy was estimated in the range of 70–80 wt%.

PPy/silver composites are composed of globules of about 1-μm diameter. This globular morphology is typical of PPy prepared with classical oxidants, such as iron(III) salts. The molecular structure of PPy produced was characterized by Fourier transform infra-red spectroscopy (FTIR) and Raman spectroscopy showing the same features as in PPy prepared with other oxidants. The morphology of silver nanoparticles was evaluated by transmission electron microscopy (TEM). Silver is present in particles of 50–100 nm size and occasionally larger polygonal crystals.

PPy/silver nanocomposites were used to coat cotton fabrics by in situ chemical oxidative polymerization using silver nitrate [39]. In a redox reaction, silver ions oxidize the pyrrole monomer and reduce to Ag^0. The reduced silver was deposited on/into the polypyrrole/cotton matrix layer as nanoparticles. In the beginning, the cotton fabric was impregnated in a solution containing pyrrole. Silver nitrate was added into this solution and stirred. After completion of reaction, the cotton fabric was coated with a PPy/silver nanocomposites layer.

The antimicrobial activity of PPy/silver-coated fabrics against *E. coli* and *S. aureus* bacteria was evaluated by the assessment test and agar diffusion test. The antimicrobial property of PPy/silver composites was measured by the clear zone of inhibition around the fabrics after incubation in agar plate method. Untreated cotton shows no antimicrobial activity against both bacteria. Moreover, the bacteria were grown over the surface. The PPy-coated cotton shows a small inhibition zone, whereas in PPy/silver composite-coated fabrics the inhibition zone was found to increase with increasing concentration of silver in the composites.

The antibacterial activity of the PPy/silver composite-coated cotton fabrics was also quantified according to the AATCC 100-1999 procedure. The PPy/silver composite-coated cotton fabrics show a gradually increased bacterial reduction percentage over the contact time. The paper reported that the bacterial reduction reaches likely 100% within 6 h against *E. coli* and 12 h against *S. aureus*.

Commercially available antimicrobial fabrics already include fabrics composed of silver-coated fibres. Therefore, another possible approach could be to treat this kind of fabrics with PPy instead to synthesize silver nanoparticles during PPy deposition as previously reported.

In a work [41], PPy deposition was carried out on cotton fabrics containing 10% of silver-coated fibres. PPy was synthesized at room temperature using ferric sulphate as oxidant. Antibacterial activity was evaluated following the AATCC Test Method 100–2012 against Gram-positive bacteria on textiles with different amount of PPy on fabrics with silver-coated fibres and pure cotton fabrics (without silver-coated fibres). A synergic biocidal effect between silver ions and PPy was observed. In fact, silver-containing fabrics used in this work alone does not guarantee a complete biocidal effect, but the addition of just 2 wt% of polypyrrole showed a bacterial reduction of 99%. On the other hand, excellent bacterial reduction (>99%) was found on pure cotton fabrics containing more than ~9 wt% of PPy, but the amount of PPy can be reduced to 5% in the presence of silver to reach the same level of efficiency.

Few papers reported the applications of PPy to man-made fibres for antimicrobial purposes. In particular, a study [42] was investigated where a polyethylene terephthalate (PET) fabric was coated with reduced graphene oxide (RGO) sheets, and then a PPy layer was deposited by in situ polymerization in order to cover RGO.

Antibacterial activity was assessed qualitatively against *S. aureus* (ATCC 25923) and *E. coli* (ATCC 25922) with AATCC 100-2003 standard method. PPy-coated sample showed antibacterial activity against both types of bacteria. The sample treated with RGO/PPy composite layer also showed excellent antibacterial activity against both bacteria that can be attributed to the existence of PPy with its antibacterial activity. No antibacterial activity was found on RGO-coated fabrics.

4. Antimicrobial agents of natural origin

Bacterial resistance to biocides, their inadequate activity, toxic effects on households and the environment and poor durability on textiles have become important issues of concern.

Some antimicrobial agents are commercially marketed as 'eco-friendly', such as Ultrafresh by Thomson Research Associates, Tinosan AM 110 (2,4,4'-Trichloro-2'-hydroxyl-diphenyl ether) by Ciba Specialty Chemicals, Sanitized AG by Clariant, Ecosy by Unitika, Utex by Nantech Textile Company Limited and Vantocil IB by Zeneca. However, investigating the chemistry behind these purportedly natural biocides, it is clear that they are not entirely natural.

As a consequence, certain synthetic antimicrobial agents, such as Triclosan, have been banned by a number of leading retailers and governments in Europe, for their potential to cause skin irritation, non-biodegradable and bioaccumulation effects [43].

Due to these concerns, coupled with the high level of consumer awareness about clothing safety, many kinds of eco-friendly antimicrobial agents such as peroxy acids, chitosan and its derivatives or specific dyes have been developed for textiles.

In the last years, significant progresses in the discovery of new compounds with antimicrobial activity, from natural products, were made. These substances may present an efficient antimicrobial effect, with safety, easy availability, non-toxicity to skin and environmental friendliness. Moreover, no resistance of pathogenic bacteria was reported towards these natural chemicals.

Plants have received interest as a major source of natural antimicrobials in nature [44, 45]. Materials extracted from different parts of plants such as bark, leaves, roots and flowers containing tannin, flavonoids and quinonoids but also alkaloids, saponins, terpenoids and phenolic compounds, with strong antimicrobial properties, have been investigated [46, 47].

Even essential oils, a mixture of a variety of aromatic compounds which can give cologne, can provide protection from a broad spectrum of microbes. The application of essential oils for antimicrobial effect on textiles has increased in recent times, due to their high efficiency, even if the real action against microbes is not clear.

A synergistic effect was noted, for example, for carvacrol and some hydrocarbon monoterpenes showing good antimicrobial properties: probably the hydrocarbons interacted with the cell membrane of the microbes and facilitated quick penetration of carvacrol into their cells. Similar effects were reported for eugenol/carvacrol and eugenol/thymol towards *E. coli*, suggesting that carvacrol and thymol disintegrated the outer membrane of *E. coli*, making it easier for eugenol to enter the cytoplasm. The advantages of synergy are the reduction of the concentration required to yield the same antimicrobial effect when compared with the sum of the purified components [48, 49].

Natural dyes, extracted from bark, leaves, roots, fruits, seeds and flowers, or from microorganisms such as fungi, algae and bacteria, have an inherent antimicrobial property due to the presence of different colouring materials such as tannin, flavonoids and quinonoids. Moreover, they offer a wide range of colours, are environmentally friendly and can be used in low-cost treatments with the additional benefit of colouring and confer antimicrobial activity in a single step [50].

Natural antimicrobial peptides, present in every living organism, are also promising natural candidates for antimicrobial textile applications. They are characterized by their small size (12–50 amino acids), the arginine and lysine residues responsible for their positive charge, and an amphipathic structure that interacts with microbial membranes. Some examples are daptomycin (Cubicin®, Cubist Pharmaceuticals), pexiganan, psoriazyna and plectasin NZ2114. Another efficient AMP is L-cysteine, successfully used to promote the biofunctionalization of wool and polyamide, conferring a durable antimicrobial finishing [51] to those fibres.

5. Chitosan

Among the antimicrobial agents of natural origin, chitosan is gaining great interest in the last decades; in fact, chitosan and its derivatives appear to be the most effective natural antimicrobial agent on the market.

5.1. Structure and properties

Chitosan (2-amino-2-deoxy-(1->4)-b-D-glucopyranan) is a natural biopolymer, resulting from the deacetylation of chitin, constituting 20–30% of the exoskeleton of crustaceans.

It is the second most abundant biopolymer in the world, following cellulose, meaning an easy availability at low cost [52]. Its natural origin makes it biodegradable, biocompatible, non-toxic and non-carcinogenic, that is, an eco-friendly product avoiding any environmental or hygienic issues.

This biopolymer shows excellent film- and coating-forming properties when cast from organic acidic water solutions and, last but not least, it shows a strong antimicrobial activity against a wide spectrum of microorganisms, including fungi, algae and some bacteria.

The antimicrobial action of chitosan is influenced by intrinsic factors and environmental conditions, such as the chitosan molecular weight and polymerization degree, its deacetylation degree, the pH of the medium and the microorganism type.

Chitosan is considered to be both bacteriocidal and bacteriostatic although the exact action mechanism is not fully understood. The most acceptable models describe the interaction between positively charged chitosan groups and negatively charged microbial cell membranes due to electrostatic interactions. It promotes changes in the properties of membrane wall permeability causing internal osmotic imbalances and consequently inhibiting the growth of microorganisms. Even the hydrolysis of the peptidoglycans in the microorganism wall occurs, leading to the leakage of intracellular electrolytes as proteins, nucleic acids and glucose. Another proposed mechanism is the binding of chitosan with microbial DNA, which leads to the inhibition of the mRNA and protein synthesis via the penetration of chitosan into the nuclei of the microorganism, reaching the plasma membrane. A third mechanism is based on the excellent metal-binding capacity of chitosan due to the amine groups which are responsible for the uptake of metal cations by chelation, suppression of spore elements and binding to essential nutrients to microbial growth [53, 54].

Due to its diversified application fields, chitosan is a biomolecule with great potential. The antimicrobial activity was undoubtedly the most interesting application in recent years, leading to a wide application of chitosan, mainly in the field of food packaging and edible films, for biomedical and pharmaceutical purposes (drug delivery or tissue engineering), cosmetics and dermatological, agriculture, paper, enzyme immobilization and, of course, in textile field.

5.2. Use as antimicrobial agent for textiles

The use of chitosan and its derivatives on fibres seems to be the more realistic prospect since this product does not provoke any immunological response.

Besides the biocidal properties of chitosan on textiles, it also has several other advantages considering the further colouration, because the amine group present readily reacts with dyes for successful dyeing/printing [55].

Chitosan is mostly applied by the traditional pad-dry-cure process using chitosan/citric acid mixture mainly on cotton fabrics, even though other techniques have been used to impart antimicrobial property to fabrics. The use of binders with chitosan has also been reported [56] with the advantage that it can be applied to all manner of fabrics due to the presence of the binder.

Complexes based on chitosan and other biocidal agents have been studied to increase treatment efficiency and durability [57, 58]. Promising results were found with bivalent metal ions, such as Cu(II), Zn(II) and Fe(II), showing an antimicrobial effect much higher than the single components, due to the stronger positive charge after complexation [59], and with nanocapsules based on antibacterial polypeptide-grafted chitosan [60].

Despite some disadvantages in the use of chitosan in textile field, namely some temperature and pH activity dependence and poor handling, Eosy®, a commercial finishing product based on this biopolymer, and a composite fibre of chitosan and viscose, named Crabyon®, presenting a durable antimicrobial efficacy, are already available [61].

In textile field, the most common way to apply chitosan to fabrics is by wet thermal curing, involving relatively high temperature with energy consumption, costs and possible fabric degradation; moreover, the addition of toxic reagents, such as glutaraldehyde, is requested as cross-linking agent.

In recent research works, ultraviolet radiation in the presence of a suitable photoinitiator was proposed as an alternative process to graft chitosan molecules to textile fibres by radical process [62, 63]. In detail, in case of a cellulosic substrate such as cotton, the same cellulose molecule can be involved in the reaction by the formation of radicals, which can react with those formed on the chitosan molecule, conferring a strong fastness to the finishing. Moreover, UV grafting is a fast and eco-friendly process, carried out at room temperature, with lower cost than the traditional thermal process.

Cotton, silk and synthetic fabrics were considered as substrates. Obtained results showed that chitosan UV curing yielded strong antimicrobial properties, reaching 100% m.o. reduction on all considered fabrics, as confirmed by antimicrobial tests carried out also on chitosan film. Moreover, low add-ons, 1–3% o.w.f., are enough to confer the desired property to the fabrics, so the hand properties of cotton or silk and the filtration capacity of synthetic fabrics are not compromised. In order to have a good treatment fastness, chitosan has to be diluted with acetic acid solution before spreading on fabrics and an impregnation time of 12 h at an ambient temperature or 1 h at 50°C is necessary before the curing to ensure a good penetration inside the fibres.

The homogeneous distribution of chitosan on fabrics was confirmed by dyeing tests with an acid dye and by SEM analysis (**Figure 2**), which showed the optimal distribution of the finish on single fibre surface, while the presence of amino groups before and after the washing test, responsible for the antimicrobial activity, was revealed by ninhydrin assay and FTIR-attenuated total reflectance (ATR) spectra.

Chitosan film was characterized by differential scanning calorimetry (DSC) and FTIR analysis. Data found are perfectly in agreement with literature data related to thermally cured

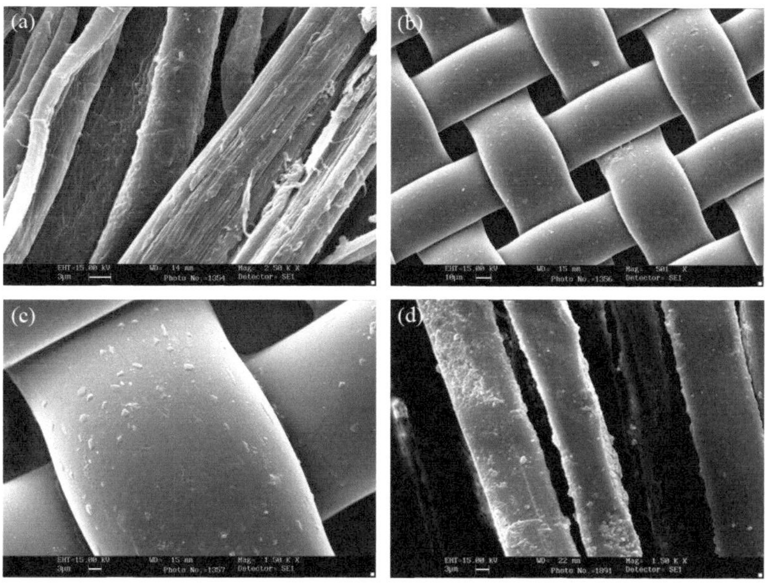

Figure 2. Chitosan UV grafted to fibres: cotton (a), PET (b), PA (c) and silk (d).

chitosan, meaning that UV curing leads at the same polymer structure. Finally, on the FTIR-ATR spectra of treated cotton or synthetic fabrics, in comparison with untreated samples, the presence of the typical bands of chitosan is evident, showing its presence again.

Chitosan was applied by UV grafting also to wool fibres, to confer a multifunctional finishing to the fabric, improving its value and application fields [64].

The antimicrobial activity, reaching 67% *E. coli* reduction, was obtained by a surface modification of wool fibres with 2% grafted chitosan, preceded by an oxidative wool pre-treatment and 1-h impregnation at 50°C to enhance chitosan penetration in wool fibres. It was coupled to an improvement of wool dyeability towards acid dyes and some anti-felting properties with respect to the untreated substrate, while the treatment fastness to laundering depended on the used surfactant and no anti-pilling properties were conferred.

A semi-industrial scale-up of the process was carried out on cotton fabrics, following an encouraging preliminary laboratory research, on samples of reduced dimensions, aimed to optimize the main process parameters. To test the feasibility of the proposed treatment at larger scale, large white or dyed fabrics were impregnated by foulard with a commercial chitosan solution, significantly reducing the add-on to restrain the costs. Then, they were irradiated, both dried and wet, with a high-power UV lamp, in air.

Obtained results confirmed the previous, laboratory scale, ones: a strong antibacterial activity with good washing fastness (99.9% microorganisms reduction after 30 washing cycles) was achieved by irradiation of the samples even wet and in air. It was obtained with a chitosan add-on percentage lowered till 0.3 wt% with a negligible affection of colour or hand properties of the fabric [65].

5.3. Wastewater purification

Wastewater treatment is one of the major current applications of chitin/chitosan-based products due to their coagulating, flocculating and chelating properties.

Ecological and health problems associated with heavy metals and pesticides accumulation in water and, as a consequence, through the food chain prompted the need for purification of industrial water in an efficient way.

The ability of the free amino groups of chitosan to form coordinate/covalent bonds with metal ions is of great interest: chitosan in the form of a film or a powder or suitably grafted to an inert substrate can be used in metal ion complexing, in particular above its pKa value (about 6.5).

Chitosan, carboxymethyl chitosan and cross-linked chitosan showed a strong efficiency in removing Cd^{2+}, Cu^{2+}, Hg^{2+}, Ni^{2+} and Zn^{2+} ions from wastewater and industrial effluents [66–69].

Chitosan was tested also as a sequestering agent for dye molecules, mainly present in wastewater from dyeing plants, showing a high efficiency towards different dye classes: acid, reactive, anionic and direct dyes [70–72].

Cotton gauzes coated with chitosan using a UV-curing process were also tested, in static and dynamic conditions, as water filter for biological disinfection against both Gram-negative and Gram-positive bacteria. The material showed good antibacterial activity against *E. coli, S.*

http://dx.doi.org/10.5772/intechopen.68423

aureus and *K. pneumoniae*, in both static assessment and dynamic conditions: chitosan-treated gauze showed a high antimicrobial efficiency in few seconds of contact time. Results are of interest even if compared with those related to the same cotton gauze cationized with a quaternary ammonium salt. A certain pH sensitivity was found, but in all cases microorganism reduction never fell under 80% [73].

It makes this composite a good candidate for its real use as biological filter.

6. Conclusions and future perspectives

The presence of Gram-positive and Gram-negative bacteria and fungi is common on textiles, involving both synthetic and natural fibres. Due to their structure and chemical composition, textile products are prone to host microorganisms whose proliferation is responsible for diseases, unpleasant odours, colour degradation and deterioration of fabrics. Many of these microorganisms are pathogens, quite often related to nosocomial infections; therefore, the development of non-toxic processes for the preparation of antimicrobial textiles is gaining interest from both the academic researches and industry.

Emerging applications for biocidal finishes in textiles are required in different fields: sportswear, underwear, textile-based medical devices, home furnishing, filtration and depuration of water or air in crowded public areas. The required effect, depending on the application, can vary from the simple odour control to disease and infection control.

Among the novel, natural and eco-friendly antibacterial finishing of textiles, a strong antibacterial finishing of textile substrates, with good fastness and stability, was obtained with both photo-grafted chitosan and polypyrrole coating on textiles.

Moreover, the morphology of PPy particles seems to effect antimicrobial performances, but the works, at the moment, were focused on particles not linked to a substrate. Further studies to improve PPy properties will appoint to produce PPy coating on the fabrics with particles having a designed shape.

In another study not dealing with textile applications, polypyrrole-graft-chitosan (PPy-g-CS) copolymer was chemically synthesized and characterized [74]. PPy-g-CS showed an antibacterial activity stronger than chitosan and PPy alone, comparable with the antibiotics considered as reference. It suggests a synergic effect of polypyrrole-chitosan coating, exploitable in textiles.

Author details

Monica Periolatto[1]*, Franco Ferrero[1], Claudia Vineis[2], Alessio Varesano[2] and Giuseppe Gozzelino[1]

*Address all correspondence to: monica.periolatto@polito.it

1 Department of Applied Science and Technology, Polytechnic of Turin, Turin, Italy

2 CNR-ISMAC, Institute for Macromolecular Studies, Biella, Italy

References

[1] Engelhardt AW. The Fiber Year 2016 [Internet]. 2016. Available from: http://www.thefiberyear.com/ [Accessed: 20 January 2017].

[2] Zanoaga M, Tanasa F. Antimicrobial reagents as functional finishing for textiles intended for biomedical applications. I. Synthetic organic compounds. Chemistry Journal of Moldova. 2014;**9**:14–32.

[3] Boryo DEA. The effect of microbes on textile material: A review on the way-out so far. The International Journal of Engineering and Science (IJES). 2013;**2**(8):09–13.

[4] Windler L, Height M, Nowack B. Comparative evaluation of antimicrobials for textile applications. Environment International. 2013;**53**:62–73.

[5] Gao Y, Cranston R. Recent advances in antimicrobial treatments of textiles. Textile Research Journal. 2008;**78**:60–72.

[6] Abdel-Mohsen AM, Abdel-Rahman RM, Hrdina R, Imramovský A, Burgert L, Aly AS. Antibacterial cotton fabrics treated with core–shell nanoparticles. International Journal of Biological Macromolecules. 2012;**50**(5):1245–1253.

[7] Mirjalili M, Zohoori S. Review for application of electrospinning and electrospun nanofibers technology in textile industry. Journal of Nanostructure in Chemistry. 2016:01–07.

[8] Sumithra M, Vasugi RN. Micro-encapsulation and nano-encapsulation of denim fabrics with herbal extracts. Indian Journal of Fibre & Textile Research. 2012;**37**:321–325.

[9] Radu CD, Parteni O, Ochiuz L. Applications of cyclodextrins in medical textiles—Review. Journal of Controlled Release. 2016;**224**:146–157.

[10] Li M, Zhang G, Xu S, Zhao C, Han M, Zhang L. Cross-linked polyelectrolyte for direct methanol fuel cells applications based on a novel sulfonated cross-linker. Journal of Power Sources. 2014;**255**(0):101–117.

[11] Harifi T, Montazer M. Past, present and future prospects of cotton crosslinking: New insight into nano particles. Carbohydrate polymers. 2012;**88**(4):1125–1140.

[12] Straccia MC, Romano I, Oliva A, Santagata G, Laurienzo P. Crosslinker effects on functional properties of alginate/N-succinyl chitosan based hydrogels. Carbohydrate Polymers. 2014;**108**(0):321–330.

[13] Dragan ES. Design and applications of interpenetrating polymer network hydrogels. A review. Chemical Engineering Journal. 2014;**243**(0):572–590.

[14] Sathianarayanan MP, Bhat NV, Kokate SS, Walunj VE. Antibacterial finish for cotton fabric from herbal products. Indian Journal of Fibre & Textile Research. 2010;**35**:50–58.

[15] Tawiah B, Badoe W, Fu S. Advances in the development of antimicrobial agents for textiles: The quest for natural products. Review. Fibres & Textiles in Eastern Europe. 2016; **24**(3(117)):136–149.

[16] Yao C, Neoh K, Shi Z-L, Kang E. Antibacterial poly(D,L-lactide)(pdlla) fibrous membranes modified with quaternary ammonium moieties. Chinese Journal of Polymer Science. 2010;**28**:581–588.

[17] Yao C, Li X, Neoh K, Shi Z, Kang E. Surface modification and antibacterial activity of electrospun polyurethane fibrous membranes with quaternary ammonium moieties. Journal of Membrane Science. 2008;**320**:259–267.

[18] Hegstad K, Langsrud S, Lunestad BT, Scheie AA, Sunde M, Yazdankhah SP. Does the wide use of quaternary ammonium compounds enhance the selection and spread of antimicrobial resistance and thus threaten our health? Microbial Drug Resistance. 2010;**16**:91–104.

[19] Simoncic B, Tomsic B. Structures of novel antimicrobial agents for textiles—A review. Textile Research Journal. 2010;**80**:1721–1737.

[20] Jones RD, Jampani HB, Newman JL, Lee AS. Triclosan: A review of effectiveness and safety in health care settings. American Journal of Infection Control. 2000;**28**:184–196.

[21] Huang C-L, Olusegun KA, Yu C-P. Triclosan: A review on systematic risk assessment and control from the perspective of substance flow analysis. Science of the Total Environment. 2016;**566–567**:771–785.

[22] Orhan M, Kut D, Gunesoglu C. Use of triclosan as antibacterial agent in textiles. Indian Journal of Fibre Textile Research. 2007;**32**:114–118.

[23] Zhao Y, Xu Z, Lin T. Barrier textiles for protection against microbes. Antimicrobial Textiles. 2016:225.

[24] Mahltig B, Zhang J, Wu L, Darko D, Wendt M, Lempa E, Haase H. Effect pigments for textile coating: A review of the broad range of advantageous functionalization. Journal of Coatings Technology and Research. 2016:1–21.

[25] Aruna S, Vasugi Raaja N, Sathiesh Kumar S. Fabrication of antimicrobial textiles using hydrothermally synthesized copper oxide nanoparticles. International Journal of Innovative Research in Science, Engineering and Technology. 2016;**5**(2):2112–2119.

[26] Raghupathi KR, Koodali RT, Manna AC. Size-dependent bacterial growth inhibition and mechanism of antibacterial activity of zinc oxide nanoparticles. Langmuir. 2011;**27**:4020–4028.

[27] Nikiforov A, Deng X, Xiong Q, Cvelbar U, DeGeyter N, Morent R, Leys C. Non-thermal plasma technology for the development of antimicrobial surfaces: A review. Journal of Physics D: Applied Physics. 2016;**49**(20):204002.

[28] Siadat SA, Mokhtari J. The role of polyhexamethylene biguanide and silver nanoparticle interaction in the fabrication of novel antibacterial bio-fibers using silk wastage. Journal of Nano Research. 2016;**43**:63–72.

[29] Zhao T, Chen Q. Halogenated phenols and polybiguanides as antimicrobial textile finishes. Antimicrobial Textiles. 2016:141.

[30] Ren X, Jiang Z, Liu Y, Li L, Fan X. N-halamines as antimicrobial textile finishes. Antimicrobial Textiles. 2016:125.

[31] Li L, Ma K, Liu Y, Liu Y, Li R, Ren X, Huang TS. Regenerability and stability of antibacterial cellulose containing triazine N-halamine. Journal of Engineered Fabrics & Fibers. 2016;**11**(1).

[32] Malinauskas A. Chemical deposition of conducting polymers. Polymer. 2001;**42**: 2957–2972.

[33] Audebert P. Recent trends in polypyrrole electrochemistry, nanostructuration, and applications. In: Cosnier S, Karyakin A, editors. Electropolymerization: Concepts, Materials and Applications. Weinheim, Germany: Wiley-VCH Verlag GmbH & Co. KGaA; 2010.

[34] da Silva Jr, FAG, Queiroz JC, Macedo ER, Fernandes AWC, Freire NB, da Costa MM, de Oliveira HP. Antibacterial behavior of polypyrrole: The influence of morphology and additives incorporation. Materials Science and Engineering C. 2016;**62**:317–322.

[35] Seshadri DT, Bhat NV. Synthesis and properties of cotton fabrics modified with polypyrrole. Sen'i Gakkaishi. 2005;**61**:104–109.

[36] Varesano A, Aluigi A, Florio L, Fabris R. Multifunctional cotton fabrics. Synthetic Metals. 2009;**159**:1082–1089.

[37] Varesano A, Vineis C, Aluigi A, Rombaldoni F, Tonetti C, Mazzuchetti G. Antibacterial efficacy of polypyrrole in textile applications. Fibers and Polymers. 2013;**14**(1):36–42.

[38] Shi Z, Zhou H, Qing X, Dai T, Lu Y. Facile fabrication and characterization of poly(tetrafluoroethylene)@polypyrrole/nano-silver composite membranes with conducting and antibacterial property. Applied Surface Science. 2012;**258**(17):6359–6365.

[39] Firoz Babu K, Dhandapani P, Maruthamuthu S, Anbu KM. One pot synthesis of polypyrrole silver nanocomposite on cotton fabrics for multifunctional property. Carbohydrate Polymer. 2012;**90**:1557.

[40] Omastová M, Mosnácková K, Fedorko P, Trchová M, Stejskal J. Polypyrrole/silver composites prepared by single-step synthesis. Synthetic Metals. 2013;**166**:57–62.

[41] Varesano A, Vineis C, Tonetti C, Mazzuchetti G, Bobba V. Antibacterial property on Gram-positive bacteria of polypyrrole-coated fabrics. Journal of Applied Polymer Science. 2015;**132**:41670.

[42] Berendjchi A, Khajavi R, Akbar Yousefi A, Yazdanshenas ME. Improved continuity of reduced graphene oxide on polyester fabric by use of polypyrrole to achieve a highly electro-conductive and flexible substrate. Applied Surface Science. 2016;**363**:264–272.

[43] Blackburn RS. Sustainable Textiles: Life Cycle and Environmental Impact. UK: Woodhead Publishing Limited; 2009.

[44] Katewaraphorn J, Aldred AK. A study of microcapsules containing *Psidium guajava* leaf extract for antibacterial agent on cotton fabric. International Journal of Chemical Engineering and Applications. 2016;**7**(1):27.

[45] Ravindra KB, Murugesh BK. Study of antimicrobial properties of fabrics treated with Ocimum sanctum L (tulsi) extract as a natural active agent. Journal of Natural Fibers. 2016;**13**(5):619–627.

[46] Hübsch Z, Van Zyl RL, Cock IE, Van Vuuren SF. Interactive antimicrobial and toxicity profiles of conventional antimicrobials with Southern African medicinal plants. South African Journal of Botany. 2014;**93**(0):185–197.

[47] Zhou Y, Yang ZY, Tang RC. Bioactive and UV protective silk materials containing baicalin—The multifunctional plant extract from Scutellaria baicalensis Georgi. Materials Science and Engineering: C. 2016;**67**:336–344.

[48] Turgis M, Vu KD, Dupont C, Lacroix M. Combined antimicrobial effect of essential oils and bacteriocins against foodborne pathogens and food spoilage bacteria. Food Research International. 2012;**48**(2):696–702.

[49] Pei RS, Zhou F, Ji BP, Xu J. Evaluation of combined antibacterial effects of eugenol, cinnamaldehyde, thymol, and carvacrol against E. coli with an improved method. Journal of Food Science. 2009;**74**(7):M379–M383.

[50] Pal A, Tripathi YC, Kumar R, Upadhyay L. Antibacterial efficacy of natural dye from Melia composita leaves and its application in sanitized and protective textiles. Journal of Pharmacy Research. 2016;**10**(4):154–159.

[51] Mouro C, Gouveia IC. Antimicrobial functionalization of wool: Assessment of the effect of Cecropin-B and [Ala5]-Tritrp7 antimicrobial peptides. The Journal of the Textile Institute. 2016;**107**(12):1575–1583.

[52] Vani R, Stanley SA. Studies on the extraction of chitin and chitosan from different aquatic organisms. Advanced Biotech. 2013;**12**:12–15.

[53] Goy RC, de Britto D, Assis BG. Polimeros: Ciencia e Tecnologia. 2009;**19**(3):241–247.

[54] Hoque J, Adhikary U, Yadav V, Samaddar S, Konai MM, Prakash RG, Haldar J. Chitosan derivatives active against multidrug-resistant bacteria and pathogenic fungi: In vivo evaluation as topical antimicrobials. Molecular Pharmaceutics. 2016;**13**(10):3578–3589.

[55] Vakili M, Rafatullah M, Salamatinia B, Abdullah AZ, Ibrahim MH, Tan KB. Application of chitosan and its derivatives as adsorbents for dye removal from water and wastewater: A review. Carbohydrate Polymers. 2014;**113**(0):115–130.

[56] Nayak R, Padhye R. Antimicrobial finishes for textiles. In: Paul R, editor. Functional Finishes for Textiles. UK: Woodhead Publishing Limited; 2015. pp. 361–385.

[57] Knittel D, Schollmeyer E. Chitosans for permanent antimicrobial finish on textiles. Lenzinger Berichte. 2006;**85**:124–130.

[58] Lim S-H, Hudson SM. Application of a fiber-reactive chitosan derivative to cotton fabric as an antimicrobial textile finish. Carbohydrate Polymers. 2004;**56**:227–234.

[59] Wang X, Du Y, Fan L, Liu H, Hu Y. Chitosan-metal complexes as antimicrobial agent: Synthesis, characterization and structure-activity study. Polymer Bulletin. 2005;**55**: 105–113.

[60] Zhou C, Wang M, Zou K, Chen J, Zhu Y, Du J. Antibacterial polypeptide-grafted chitosan-based nanocapsules as an "armed" carrier of anticancer and antiepileptic drugs. ACS Macro Letters. 2013;**2**:1021–1025.

[61] Santos Morais D, Guedes RM, Lopes MA. Antimicrobial approaches for textiles: From research to market. Materials. 2016;**9**:498.

[62] Ferrero F, Periolatto M. Antimicrobial finish of textiles by chitosan UV curing. Journal of Nanoscience and Nanotechnology. 2012;**12**:4803–4810.

[63] Periolatto M, Ferrero F, Vineis C. Antimicrobial chitosan finish of cotton and silk fabrics by UV curing with 2-hydroxy-2-methylphenylpropane-1-one. Carbohydrate Polymers. 2012;**88**:201–205.

[64] Periolatto M, Vineis C, Ferrero F, Rombaldoni F. Multifunctional finishing of wool fabrics by chitosan UV-grafting: An approach. Carbohydrate Polymers. 2013;**98**:624–629.

[65] Ferrero F, Periolatto M, Ferrario S. Sustainable antimicrobial finishing of cotton fabrics by chitosan UV-grafting: From laboratory experiments to semi industrial scale-up. Journal of Cleaner Production. 2015;**96**:244–252.

[66] Fernandez-Saiz P. Jose′ M. Lagaron, Maria J. Ocio, Amparo Lopez-Rubio Chitosan and chitosan blends as antimicrobials. In: Antimicrobial Polymers. John Wiley and Sons; 2012. pp. 71–91.

[67] Ferrero F, Tonetti C, Periolatto M. Adsorption of chromate and cupric ions onto chitosan-coated cotton gauze. Carbohydrate Polymers. 2014;**110**:367–373.

[68] Vunain E, Mishra AK, Mamba BB. Dendrimers, mesoporous silicas and chitosan-based nanosorbents for the removal of heavy-metal ions: A review. Int J Biol Macromol. 2016;**86**:570–586. DOI: 10.1016/j.ijbiomac.2016.02.005.

[69] Zhang L, Zeng Y, Cheng Z. Removal of heavy metal ions using chitosan and modified chitosan: A review. Journal of Molecular Liquids. 2016;**214**:175–191.

[70] Dotto GL, Santos JMN, Tanabe EH, Bertuol DA, Foletto EL, Lima EC, Pavan FA. Chitosan/polyamide nanofibers prepared by Forcespinning® technology: A new adsorbent to remove anionic dyes from aqueous solutions. Journal of Cleaner Production. 2017; **144**:120–129.

[71] Gonçalves JO, Santos JP, Rios EC, Crispim MM, Dotto GL, Pinto LAA. Development of chitosan based hybrid hydrogels for dyes removal from aqueous binary system. Journal of Molecular Liquids. 2017;**225**:265–270.

[72] Periolatto M, Ferrero F. Cotton filter fabrics functionalization by chitosan UV grafting for removal of dyes. Chemical Engineering Transactions. 2013;**32**:85–90.

[73] Ferrero F, Periolatto M, Vineis C, Varesano A. Chitosan coated cotton gauze for antibacterial water filtration. Carbohydrate Polymers. 2014;**103**:207–212.

[74] Cabuk M, Alan Y, Yavuz M, Unal HI. Synthesis, characterization and antimicrobial activity of biodegradable conducting polypyrrole-graft-chitosan copolymer. Applied Surface Science. 2014;**318**:168–175.

Chapter 3

Inorganic Nanoparticles: Innovative Tools for Antimicrobial Agents

Mario Kurtjak, Nemanja Aničić and
Marija Vukomanovicć

Additional information is available at the end of the chapter

http://dx.doi.org/10.5772/67904

Abstract

Resistance of bacteria to antibiotics is an urgent problem of humanity, which leads to a lack of therapy for serious bacterial infections. Development of new antibiotics has almost ceased in the last decades—even when a new antibiotic is launched, very soon the resistance of bacteria appears. There is a long list of applications where antimicrobial protection is required to achieve effective treatment. However, if we use the same antibiotics for all these applications, we will remain caught in the "vicious circle" of constant discovery of new synthetic antibiotics and very fast development of their resistant species. Therefore, we need to find alternative strategies that will be routinely used for some specific conditions (wounds, implants, etc.). Thus, we will keep the activity of antibiotics and save them for acute conditions (pneumonia, meningitis, etc.). An option for designing alternative antimicrobial strategies is to go back to the antimicrobials that were used before the discovery of antibiotics, i.e., inorganic antimicrobial agents including ions (Ag^+, Cu^+/Cu^{2+}, Zn^{2+}, Ga^{3+}, etc.) or nanoparticles (Ag/AgO, $Cu/Cu_2O/CuO$, ZnO, Ga/Ga_2O_3, TiO_2, MgO, V_2O_3, etc.). Here we are going to summarize the main properties of inorganic antimicrobials as well as advantages, disadvantages and perspectives for their application.

Keywords: Ag/AgO, $Cu/Cu_2O/CuO$, ZnO, Ga/Ga_2O_3, TiO_2, MgO, V_2O_3, functionalized Au

1. Introduction

Resistance of bacteria to antibiotics is becoming an increasingly urgent problem of the humanity. The most serious threat comes from vancomycin-resistant *Enterococcus* (VRE, mainly *E. faecium*), methicilin-resistant *Staphylococcus aureus* (MRSA), *Klebsiella* (especially *K. pneumoniae*), *Acinetobacter baumanii, Pseudomonas aeruginosa, Enterobacter* and *Escherichia coli* (the so-called "ESKAPE" pathogens,), Gram-positive *Mycobacterium tuberculosis* and some other Gram-negative bacteria [1]. Soon there will be no available antibiotics to treat infections with these pathogens. The problem first appeared in hospitals and grew promptly as a consequence of uncontrolled application of antibiotics not only in the healthcare but also in agriculture, stock breeding, poultry breeding, etc. However, overuse and misuse are not the only factors that speed up the spread of resistance. Some mechanisms of resistance do not destroy the antibiotic and leave it active in the environment. Thus, bacteria themselves help maintain the antibiotic environment; furthermore, the drug can be released into other environments and alter them. Many precautions against drug misuse and overuse led to the reduction of antibiotic application in the last decade. Consequently, the spreading of resistance slowed down, but it did not decrease. We could get rid of the resistant strains with new antibiotics. Unfortunately, development of new antibiotics has almost ceased in the last decades. Investments in research and development of new kinds of antibiotics were minimized due to their unprofitability. And even when a new antibiotic is launched, very soon the resistance of bacteria to the new antibiotic appears.

What can we deduce from all these facts? Instead of focusing only on development of new antibiotics, which will sooner or later create resistance, we should focus on preventing the resistance itself. There is a long list of applications where antimicrobial protection is required in order to achieve effective treatment. However, if we use the same antibiotics for all these applications, we will remain caught in the "vicious circle" of constant discovery of new synthetic antibiotics and very fast development of their resistant species. Therefore, we need to find alternative strategies that will be routinely used for some specific conditions (such as insufficient and slow wound healing, rejection of medical implants during their incorporation into the body due to the presence of bacteria on the surface of the implant, unsuccessful use of autologous, allogeneic or xenografts in tissue engineering because of the development of infection, etc.). Thus, we will keep the activity of the antibiotics and save them for urgent, acute conditions (like pneumonia, meningitis, peritonitis, etc.). One option for designing these alternative antimicrobial strategies is to go back to the antimicrobials that were used before the discovery of antibiotics, i.e., inorganic antimicrobial agents. There are a lot of inorganic substances with the capacity to kill bacteria or to inhibit bacterial growth. They are applicable in the form of antibacterial ions (i.e. Ag^+, Cu^+/Cu^{2+}, Zn^{2+}, Ga^{3+}, etc.) or antibacterial nanoparticles (i.e. Ag/AgO, $Cu/Cu_2O/CuO$, ZnO, Ga/Ga_2O_3, TiO_2, MgO, V_2O_3, functionalized Au, etc.). The new knowledge brought especially by the emergence and progress of biomaterials science and nanotechnology might enable: (i) local, targeted action without side effects in the organism, (ii) improved transport towards and eased penetration into the pathogenic species, leading to higher efficiency, (iii) unique opportunity for development of effective medicines.

This chapter provides detailed overview of various inorganic antimicrobial agents, their physicochemical properties and various mechanisms of action on bacterial/mammalian cells.

2. Antibacterial ions

2.1. Silver (I) (Ag^+)

Silver nitrate ($AgNO_3$) was widely used for treatment of ulcers, burn wounds and different infections until the discovery of penicillin and sulpha drugs completely drove it out from the market [2]. In 1965, the favourability of $AgNO_3$ over antibiotics was shown and burn treatment procedure that involved frequent wetting of a cotton gauze dressing with 0.5 wt.% $AgNO_3$ solution was established [2, 3]. In spite of reduced mortality from severe burns and strong action against *Staphylococcus aureus*, haemolytic streptococci, *Pseudomonas aeruginosa* and *Escherichia coli*, the 0.5% $AgNO_3$ method had several disadvantages. It required very clean wounds, so deep and large burns were prone to invasive infection (usually by *P. aeruginosa*) leading to sepsis and death. 0.5% $AgNO_3$ was hypotonic, sensitive to light, inactive against *Aerobacter*, *Paracolon*, *Klebsiella* and a number of cutaneous saprophytes and could cause methaemoglobinaemia. Although the precipitation of $AgNO_3$ with different anions resulted in low absorption of silver into the body through the wounds, it consequently caused Na^+, K^+, Ca^{2+}, Mg^{2+} and Cl^- depletion in serum [2–4]. Hence, an improvement was tried by combining $AgNO_3$ with a sulpha drug to obtain silver sulphadiazine [2]. Many other ionic Ag drugs emerged, but Ag-sulfadiazine remained the most widely used, although it delays the wound healing process [2, 5]. Further development went to systems for controlled delivery of Ag^+ ions, Ag-containing wound dressings, catheters and antibacterial coatings [6, 7] which flooded the market recently [8].

The antibacterial action of Ag^+ ions is currently explained by three mechanisms:

1. Ag^+ ions react with thiol groups of the respiratory and transport proteins in the cell membrane [6, 9] so that cellular respiration and electron transfer are blocked [6, 9], membrane potential and permeability are disrupted, leading to cell death [10].
2. Ag^+ ions enter the bacterial cells either through ion channels or due to the detachment of the cytoplasm membrane [9, 11]. Once inside, they complex with nucleobases of DNA and RNA leading to DNA condensation and loss of replication ability [6, 9, 12].
3. Increased production of reactive oxygen species (ROS). Disruption of cellular respiration and inactivation of intracellular thiol-based antioxidants increases the oxidative stress caused by reactive radicals that are generated by the Fenton reactions [9, 13].

There are two forms of resistance: complexation of Ag^+ inside cells or reduced permeability to Ag^+ combined with an upgraded active efflux mechanism to pump Ag out of the cell [5]. Several Gram-negative and positive bacteria have been reported to be Ag-resistant, including *Pseudomonas aeruginosa*, *Pseudomonas stutzeri*, members of the *Enterobacteriaceae* and *Citrobacter spp*. [6]. It has been recently shown that the toxic values of $AgNO_3$ and commercial Ag-containing wound dressings for *P. aeruginosa*, *E. coli*, *S. aureus* and human fibroblasts are very similar and that thiol-containing molecules reduce their toxicity towards both prokaryotes and eukaryotes [14].

2.2. Copper (I, II) (Cu^{+}, Cu^{2+})

$Cu^{+/2+}$ has also been known as a sterilizing, antiseptic and antimicrobial agent [15] used to treat a variety of skin diseases, syphilis, tuberculosis and anaemia, and to fight mildew [10, 16, 17]. In modern healthcare, the antimicrobial effect of Cu is very effectively used in hospital water distribution systems [16, 17]. Recent research has focused on "contact killing" mechanism [17]. In 2008, the US Environmental Protection Agency (EPA) proclaimed Cu-surfaces as efficient antimicrobials [17]. Cu is the first metal to be awarded such a status [17]. Cu is an essential micronutrient and more than 30 types of proteins that contain Cu ions are known today [18]. In these enzymes, Cu serves as an electron donor/acceptor by alternating between the redox states Cu(I) and Cu(II) [19]. Dietary intakes of 0.9–1.4 mg of Cu for an adult (a 70-kg person) and 50 μg per kg body weight per day in infants are recommended by the WHO [16, 20]. Cu ions are also toxic to prokaryotes and eukaryotes at higher cellular concentrations, and the involvement of Cu (and Zn) in phagosomal killing of bacteria engulfed by macrophages is an important defence mechanism [10, 21].

The antibacterial/toxicity action of Cu(I, II) is currently explained by the following mechanisms:

1. Direct generation of ROS through Fenton-type reactions [19, 22]. Radicals can cause oxidative damage to proteins, nucleic acids and lipids, which lead to cell death [23].
2. Indirect generation of reactive oxygen species by inactivation of antioxidants and thiol depletion [19, 23]. Such reactions of Cu can lead to the inhibition of respiratory enzyme function and disruption of respiration leads to ROS as explained for Ag^{+} [6].
3. Competition with other metal ions for important binding sites on proteins [6, 17, 19]. Site-specific inactivation by Cu ions can also occur in Fe-S dehydratases, the cytoplasmic enzymes needed to make branched-chain amino acids [17, 23].
4. By cross-linking within and between strands of DNA, Cu may cause helical structure disorders and DNA denaturation [16, 24]. Some studies have shown that H_2O_2 was required for the DNA breakage, which questions the relevance of this mechanism [16].

Bacteria have evolved a range of mechanisms to protect themselves from the toxic effects of excess Cu ions: exclusion by a permeability barrier; intra- and extracellular sequestration of Cu ions by cell envelopes and metallothionein-like Cu-scavenging proteins in the cytoplasm and periplasm; active transport membrane efflux pumps; reduction in the sensitivity of cellular targets to Cu ions; extracellular chelation or precipitation by secreted metabolites including Cu; and adaptation and tolerance via up-regulation of necessary genes in the presence of Cu [16, 19, 25]. Active extrusion of Cu from the cell appears to be the chief mechanism of Cu tolerance in bacteria and has been extensively studied in Gram-positive and Gram-negative bacteria. However, due to the multiple targets and mostly non-specific mechanisms of damage exerted by Cu, this bacterial tolerance is relatively low, as compared to the resistance to antibiotics (i.e., 10-fold lower sensitivity to Cu as opposed to 1000-fold less sensitivity to methicillin, for example, by methicillin-resistant *S. aureus*).

2.3. Zinc(II) (Zn^{2+})

Zn^{2+} is also an essential micronutrient for the development, growth and differentiation of all living systems, including bacteria, and exhibits antibacterial action only at higher concentrations when its homeostasis is overcome. The adult human body contains approximately 1.5–2.5 g of Zn^{2+} [22, 26–28] with essential role in cell membrane integrity, development and maintenance of the body's immune system, managing insulin action and blood glucose concentration, bone and teeth mineralization, normal taste and wound healing [22]. Zn is a constituent of more than 300 enzymes that have a central role in reconstruction of the wound matrix [26, 29]. Zn in castor oil has a special place in the treatment of nappy (diaper) rash [26]. A vast range of zincated bandages, dressings, emollients, shampoos and creams are available commercially. In normal wound healing, body creates a higher amount of Zn^{2+} in the wound margin at a certain stage—during the formation of granulation tissue, scar tissue and re-epithelialization. It is believed that the addition of Zn at this stage might accelerate wound healing. Experimental studies have shown that topical ZnO reduced the initial haemorrhagic phase and promoted the regrowth of damaged skin and hair [26]. The antibacterial properties of Zn^{2+} ions are exploited especially in oral healthcare for prevention of caries, gingivitis and periodontitis. Zn− salts are used in mouthwashes and toothpastes [30]. The effect of Zn^{2+} ions is most probably only bacteriostatic, so oral-care products are designed for frequent use, while bactericidal action can be obtained in combinations with fluoride or Triclosan [30–33].

The antibacterial action of Zn− ions is a consequence of the following mechanisms [6, 30–32]:

1. Inhibition of enzymes that contain sulfhydryl groups and require Mg^{2+} ions; competitive inhibition and reaction of Zn(II) with sulfhydryl groups [30–32].

2. Zn^{2+} ions inhibit the utilization of the bacterial carbon source. They can disrupt the metabolism of sugars as well as the amino acid metabolism.

3. Zn^{2+} reduces the acid tolerance of *S. mutans* by inhibiting the transmembrane proton-translocating F-ATPase, which is the main engine for acid tolerance [30].

4. Zn(II) binds to the membranes and slows down the growth of organisms [6], inhibits protease-induced adhesion [34] and reduces the net negative charge on the cell surface and, hence, increases co-aggregation [34].

Resistance of bacteria to toxic levels of Zn^{2+} can be due to extracellular accumulation, sequestration by metallothioneins, intracellular physical sequestration, and/or can be efflux based [35]. A recent study compared the Cu and Zn resistance of MRSA and methicillin-susceptible *S. aureus* in a global collection of species [36]. While there was no difference in their Cu− susceptibility, there were significantly more Zn-resistant MRSA strains, which also had an encoded Zn resistance [36]. Similarly to Ag, recent progress of Zn− antimicrobials has gone in the direction of ZnO nanoparticles and incorporation of ionic Zn into zeolites, polymers, bioactive ceramics and glasses to achieve better efficiency and local action [6].

2.4. Gallium(III) (Ga^{3+})

Antibacterial properties of Ga^{3+} were first mentioned in 1931 [37]. Initially, it was mainly investigated for cancer diagnosis and treatment [38, 39]. Intensive research of Ga(III) as an antibacterial agent in the 2000s revealed great efficacy against *M. tuberculosis* [40] and *P. aeruginosa* [41, 42]. A recent study has shown that $Ga(NO_3)_3$ at safe therapeutic dosage (10 mg/kg) protects mice from *M. tuberculosis* infection [43]. A Phase-1 clinical study is being conducted since 2010, which tests Ganite in human patients suffering from cystic fibrosis, and chronically infected by *P. aeruginosa* [37, 38, 44]. Current results show that intravenous Ganite infusion for 5 days decreases the amount of *P. aeruginosa* in the lung without any serious adverse effect [38, 44]. Subcutaneous application of Ga-maltolate was effective in reducing *S. aureus, A. baumannii* and *P. aeruginosa* colonization in burn wounds of thermally injured mouse model [42]. These data support a potential use of Ga-maltolate *in vivo*, especially in topical administration for the prevention and treatment of wound infections. Besides $Ga(NO_3)_3$ and Ga-maltolate, some other forms of Ga(III) have also been used, i.e., chloride [45], citrate [46], desferriox-amine B and other complexes [46–48].

The following is currently known about the mechanism of antibacterial action of Ga^{3+} ions:

1. Ga^{3+} follows uptake and transport pathways for Fe^{3+}; unlike Fe(III), it cannot be reduced to the oxidation state (+2); small amounts of non-bound Ga can exist in solution at physiological conditions, versus insignificant amounts of non-bound Fe^{3+}, permitting biological interactions for Ga^{3+} that would not be possible for Fe^{3+} [49, 50].

2. Most bacteria require Fe for growth [37]. If bacteria use Ga instead of Fe, it will prevent their multiplication, which is crucial for harming the organism (as observed in bacterial ferric-binding protein and non-ribosomal peptide microbial siderophores [49]).

3. Ga(III) can affect the synthesis of siderophores by regulation of gene expression [68] leading to shortage of Fe inside cell and inhibition of many Fe-requiring enzymes.

4. Increased production of H_2O_2 was noticed due to Ga^{3+} antibacterial action [51]. However, Ga^{3+} quenches the superoxide ion signal [51], and it is not yet clear whether the ROSs are the main reason or only a consequence of the Ga^{3+} antibacterial action.

Considerable progress has been recently made in the development of Ga delivery systems using phosphate-based glasses [52–54], cellulose [55], scaffolds [56], phosphosilicates [57, 58] and titanium implants [59]. Because bacteria cannot discriminate between Fe(III) and Ga(III), they will not sense an increase of Ga^{3+} concentration and a decrease of Fe^{3+}. However, since Ga(III) enters microbial cells by exploiting specific Fe(III)-uptake mechanisms, mutations in these pathways could block Ga from reaching its cellular targets, ultimately making bacteria less susceptible to Ga's inhibitory activity, as it has been observed in laboratory studies of Ga(III) antibacterial mechanism, in which resistant strains were created by genetic modification of *P. aeruginosa* [60, 61]. Nevertheless, such mutations could never completely prevent Ga^{3+} entrance into bacterial cells and only 2–4 times higher Ga(III) concentrations were already effective against the resistant strains.

http://dx.doi.org/10.5772/67904

3. Antibacterial nanoparticles

3.1. Ag nanoparticles

Ag nanoparticles show bactericidal action in both Gram-positive and Gram-negative bacteria, with higher efficiency induced by smaller particles [12, 32, 62] and quite intriguing dependence of the efficiency on the shape, falling in the order: triangular nanoplates, nanospheres, nanowires [63]. In Ag nanoparticles, there are three sources of bactericidal activity: Ag, Ag ions and nanosize. As the three sources are interlacing, it is difficult to determine what effect comes from each of them. Ag nanoparticles come into contact with bacterial cells. Positively charged Ag nanoparticles attach to bacterial membrane by electrostatic interactions, while negatively charged ones attach due to high affinity of Ag (soft acid) for P- and S-containing molecules (soft bases) [63–65]. Then, Ag ions are released into the cell and inhibit respiratory enzymes, which facilitates the generation of ROS and consequently damages the cell membrane [66]. The uptake of Ag can be recognized by irregular pits. They can be dissolved by oxygen or H_2O_2. A Fenton-like reaction was suggested to account for the observed generation of OH• radicals at pH below 7.4 [67]. Then, S-containing proteins in the membrane or inside the cells and P-containing elements like DNA are likely to be the preferential sites for Ag nanoparticle binding. Disruption of membrane morphology may cause a significant increase in permeability, leading to leaking of the internal components resulting in cell death [63].

Exposure of murine macrophages to Ag NPs showed mitochondrial damage, apoptosis and cell death abrogated in the presence of Ag ion-reactive, thiol-containing compounds suggesting the central role of Ag ions in Ag NP toxicity [68]. Further research showed that Ag^+ ions were the only active part of the Ag NPs [69]. Testing under anaerobic conditions (ion release was negligible) showed that Ag NPs were ineffective against *E. coli* K12. If the Ag NPs were exposed to air prior to the antibacterial test under anaerobic conditions, their antibacterial properties were enhanced and bacterial survivability depended on released ions.

However, the amount of released ions from the Ag NPs at their MIC was always lower than the MIC of Ag^+ ions [70–72]. Recent research [73, 74] showed much higher intracellular dissolution of Ag NPs compared to extracellular ones. The ion release from the Ag NPs is size-specific and surface-dependent. The toxicity of 20–80-nm Ag NPs follows this size dependence and is mainly assigned to the released ions. However, the 10-nm Ag NPs are much more toxic. Importantly, immobilized Ag NPs were more efficient than Ag^+ ion-releasing substrates, even though they released much lower amount of ions and the immobilized Ag NPs were not internalized [72, 75]. Ag NPs can change the lipid composition of the membrane, anchor and incorporate into the outer membrane, and it is currently believed that the outer membrane damage is mainly "nano-specific" [72]. Ag NPs enhance the transport of Ag^+ ions into the cell and could avoid bacterial resistance that involves efflux systems. However, *E. coli* easily develop resistance to Ag NPs as well as Ag^+ after 100–200 generations of exposure to Ag NPs [76] and several studies have shown low efficiency of Ag NPs against Ag-resistant bacteria [71, 72].

3.2. Cu/CuO nanoparticles

In Cu nanoparticles, there is a coincidence of antibacterial effect of ions and nano-sized particles. The efficiency of Cu was improved by decreasing the dimensions, but it was higher for Gram-positive bacteria [32]. Cu nanoparticles have great affinity for amines and carboxyl groups, so they bind to the ones on the surface of bacteria and release the ions inside. These ions can then interact with DNA molecules and intercalate with nucleic acid strands [77]. It is believed that here, the role of ROS is much larger than in Ag nanoparticles, since they can be generated by CuO as well as the released Cu^{+}/Cu^{2+} ions by their dissolution [78]. Some scientists, on the other hand, emphasize the role of the released ions more [32]. Both Cu and CuO antibacterial nanoparticles cause lipid peroxidation, cell wall and membrane damage and oxidative damage to DNA. They generate ROS in the absence of any cells, in extracellular as well as intracellular environment. CuO nanoparticles are much more toxic to mammalian cells than Cu^{2+} ions and also much more cytotoxic than ZnO and TiO_2 NPs [79]. In general, it has been shown that trends in bactericidal activity were similar to trends in cytotoxicity, i.e. more powerful bactericidal agents [80] were more toxic towards human cells [81].

3.3. ZnO nanoparticles

ZnO has so far been found to be the most effective metal oxide antimicrobial, with efficiency comparable to Ag [32]. If ZnO nanoparticles are shined with UV light, their antibacterial effect can become strongly bactericidal as a consequence of photocatalysis. ZnO is a semiconductor with a direct 3.3-eV band gap [82]. Absorption of light with energy greater than 3.3 eV induces the electron transfer from the valence to the conduction band and separation of charge, generating a hole (h^+) in the valence band and an electron (e^-) in the conduction band [82, 83]. At the surface of the excited ZnO particle, the valence band holes abstract electrons from water and/or hydroxide ions, generating hydroxyl radicals (OH•). Electrons can reduce O_2 to produce the superoxide anion O–•. The obtained OH• and O–• can induce lipid peroxidation in membranes, DNA damage due to strand breakage or oxidized nucleotides and oxidation of amino acids and protein catalytic centres [83]. Negative charge of OH• and O–• prevents these species from passing through the membrane into the cell, so they can exert only outside damage. They can also combine with H^+ to create H_2O_2, which can pass into the cell and create internal damage leading to cell death [82]. ZnO nanoparticles show bactericidal properties as well as ROS generation also in complete absence of light. This effect has been tried to be explained by surface defects and the oxidative role of oxygen or halogens adsorbed on their surfaces [31, 82]. Such a mechanism would be enhanced in an aerobic environment and it was observed that oxygen annealing and formation of nanoholes on the surface, which both stimulated a high amount of adsorbed oxygen atoms on the ZnO surface, increased the ROS production and enhanced the antibacterial properties [31, 82]. ZnO nanorods are stronger antimicrobials than nanospheres, and flower-shaped nanoparticles with exposed polar are even stronger [82].

3.4. TiO_2 nanoparticles

TiO_2 nanoparticles can account for their antibacterial effect as a consequence of production of OH• radicals. They do not possess any antibacterial properties in the absence of UV light due

http://dx.doi.org/10.5772/67904

to weak interaction with bacterial surface because of negative charge. In contrast, recent studies have shown that TiO_2 particles exhibited high tendency to bond to the *E. coli* membrane via Van der Walls and receptor-ligand interactions. The extent of these interactions was more pronounced than in the case of ZnO nanoparticles. As a result, when illuminated with UV light, TiO_2 was more powerful than ZnO. As opposed to TiO_2, it did not show up-regulation of ROS-related proteins but rather caused membrane damage via direct transfer of ROS molecules from particle surface towards the bacterial membrane [84]. Of interest is metal doping (e.g. with Ag) of TiO_2, which can improve its antibacterial properties significantly and enable visible-light-induced photocatalytic activity [85, 86].

3.5. Functionalized Au nanoparticles

Au nanoparticles alone are considered biocompatible and bioinert [87–89]. Only Au NPs with size below 3 nm are cytotoxic due to their irreversible binding to key biopolymers [90]. Internalization of Au NPs into a cell is size-, shape- and charge-dependant. The fastest uptake was observed for 40–50 nm size; it was higher for nanospheres vs nanorods and positively charged NPs penetrate more easily [91–93]. Au NPs can be used as antibacterial agents only if they are irradiated with NIR light (photothermal treatment) or if some antibacterial component is added to them [77, 85, 89]. Interestingly, some studies have also shown antibacterial activity of Au nanoparticles with non-antibacterial components added to them, like C/Au core shell [94] or functionalized Au nanoparticles [95–97]. Au nanoparticles as carriers enable entry of the added molecules into bacterial cells, where they can directly affect some important molecules, otherwise protected by the cell wall and membrane. Concentration of otherwise inactive molecules on the surface of Au nanoparticle enables (or increases) some interactions that lead to bacterial death [95]. In this way, 4,6-diamino-2-pyrimidinethiol was able to chelate Mg^{2+} ions when attached to the Au nanoparticle [95] and induced damage of the outer membrane, leading to increased permeability of the cellular membrane. Nanoparticles entered the cell, where chelation of Mg^{2+} and interaction of the particles with DNA resulted in inhibition of protein synthesis. Cell death followed as a consequence of leakage of intracellular contents [95]. The antibacterial action of Au/4,6-Diamino-2-pyrimidinethiol NPs involves changing the membrane potential and inhibition of ATP synthase activities to decrease the ATP level and inhibition of the ribosome subunit for tRNA binding, indicating a collapse of biological process [98]. Alternatively, in amino acid-functionalized Au NPs, a structure similar to antimicrobial peptides was created and enabled strong electrostatic interactions between cationic functionalization at Au NPs and bacterial membrane resulting in damage of the membrane compactness and structure which provided antibacterial action in *E. coli* and *S. aureus* [97].

3.6. Gallium-containing nanoparticles

Investigation of Ga-based antibacterial nanoparticles has begun only very recently and it started with Ga_2O_3 nanoparticles (100 nm) showing their anti-biofouling properties against *E. coli* and *S. aureus* [99]. However, concentrations up to 25 mg/L (133 μM) exhibited only very weak (towards *S. aureus*) or no inhibition (towards *E. coli*) of planktonic growth. Further investigation showed antibacterial action of Ga_2O_3 nanorods (50×200 nm) which created an inhibition zone in *E. coli* already at 25 mg/L concentration, whereas at least 50 mg/L concentration was needed for an inhibition zone in *S. aureus* [100]. By contrast, bulk Ga_2O_3 did not create any inhibition zone. They also presented good photocatalytic properties of this semiconductor with a band gap of 4.9 eV, but photocatalysis was not responsible for the observed antibacterial action, since the test was performed in dark and Ga_2O_3 nanoparticles can create reactive oxygen species (only OH• radicals) only under direct visible light illumination [101]. Antibacterial activity was shown also for GaN nanoparticles (50 nm) [102] and anti-biofouling activity was observed towards *S. aureus, E. coli, P. aeruginosa* and *Pseudomonas putida*. Another study about a new Ga(III) delivery system in the form of $KGa[Fe(CN)_6]$/PVP NPs (average size of 15 nm) was published, which demonstrated their very good biocompatibility with HeLa cells until at least 1.1 mM concentration, their most probable endocytotic penetration into the cells and untargeted distribution in the cytoplasm, and *in vitro* exchange of Ga(III) by Fe(III) from Fe(II) suggesting their ability to sequester Fe(II) and consequently release Ga(III) [103]. A very recent study on eutectic GaIn alloy nanoparticles (average size around 100 nm) has also shown their low *in vitro* cytotoxicity against HeLa cells for at least 21 mg/L (0.2 mM Ga) concentration and endocytosis, fusion and degradation of the eutectic GaIn nanoparticles with release of Ga^{3+} ions inside HeLa cells [104]. The *in vivo* injection of these nanoparticles into mice caused no tissue damage, no allergic reaction, exhibited very low acute toxicity (maximum tolerated dose of 700 mg/kg), while Ga and In were excreted with both faeces and urine [104]. However, the antibacterial properties of KGa[Fe(CN)6]/PVP and eutectic GaIn alloy nanoparticles have not been evaluated. On the other hand, Narayanasamy et al. incorporated Ga(III)-tetraphenyl porphyrin into polymer nanoparticles (average size of 300 nm) and demonstrated their efficiency against *Mycobacterium smegmatis* as well as against HIV in macrophages, and did not show any sign of cytotoxicity for macrophages even at 2 mM concentrations despite internalization of the nanoparticles into all compartments of the cells [105]. Another way for the local delivery of Ga(III)-tetraphenyl porphyrin was by its conjugation to Pt nanoparticles (average size around 30 nm) [106]. Bactericidal properties against *S. aureus* were demonstrated under visual light illumination. However, the non-conjugated Pt nanoparticles were not tested, so it is not clear how large their contribution was and to what extent they were only deliverers of Ga(III)-tetraphenyl porphyrin. First investigation on antibacterial performances of elemental Ga nanoparticles [107] confirmed activity against *P. aeruginosa* with MIC at 0.1 mg/ml, low toxicity at this concentration and wide therapeutic window, which gives a good promise to this material for further investigations and design for biomedical applications.

3.7. Nanostructured MgO

MgO exhibits a broad range of antimicrobial activities against both Gram-positive and Gram-negative bacteria comparable to ZnO [107]. The molecular mechanism of MgO's antibacterial activity is still unclear. Some reports show dominant role of ROS which cause lipid peroxidation and diffuse inside the cell and cause cell death [108–110]. Other reports show non-ROS mechanism and suggest polar interaction with components of the cell wall (e.g. lipopolysaccharides), which cause membrane disintegration and bacterial death [111] similar to antimicrobial peptides [112]. In both mechanisms' descriptions, the surface defect sites were related to the production of ROS. In the latter case, generation of ROS species was attributed to defects in general [111]. In the former case, oxygen could be reduced at the surface oxygen vacancy [109]. This is consistent with the mechanism of ROS species generation at the surface of MgO. However, for this to happen, energy is required to support electron transfer from vacancy towards the molecular oxygen [113]. Other mechanism of ROS generation was completely ignored in explanation of MgO antibacterial activity [113]. Till date, the following properties of MgO are known: (i) MgO exhibits contact-based antibacterial action [108]; (ii) increasing the pH in bacterial suspension due to MgO hydration did not contribute to its antibacterial activity [108]; (iii) dissolved Mg^{2+} were not causing harm to bacteria [108]; (iv) AFM and SEM morphology studies confirmed deterioration of bacterial membrane, which indicated membrane leakage [111, 114]; (v) TEM study showed lack of MgO particle internalization in bacteria, which indicated that MgO particles are "doing the damage" outside of the bacteria [111]. It has been shown that the bactericidal potential of MgO is proportional to its specific surface area. The MgO with the highest specific surface area (BET) exhibited the most effective antibacterial activity [115]. Improvement in the effectiveness of the bacteria/surface contact is also achieved by Li-doping which enhanced the creation of oxygen vacancies and improved antibacterial activity [116]. The strength of the nanoparticle interaction is inversely proportional to the size, i.e. smaller particles exhibited stronger agglomeration [117]. The agglomeration could strongly influence the further processing of MgO nanoparticles [118]. MgO particles containing microrods exhibited moderate antibacterial activity, while nano-textured microrods showed strongly improved antibacterial activity. As-prepared particles exhibited reduced agglomeration, lower specific surface area and improved bactericidal potential when compared to the commercial MgO nanoparticles. We attributed the difference in antibacterial activity to a reduced concentration of non-emissive defects at the surface of nano-textured MgO microrods [118]. Magnesium is the second most abundant intracellular cation in the human body [119] essential in many physiological processes like enzyme activity, membrane processes, functioning of muscle and neural tissue, and so on. [119]. The clinical study showed the ability of MgO to reduce hypertension (1g for 21 days) [120]. Although *in vitro* studies pointed out toxic effect of MgO on human cells [121], at a concentration of 0.2 mg/ml in suspension MgO particles were able to eliminate bacteria while at the same time showed potential to exhibit bioactive properties on the cells. In this context, there is a possibility to exploit multifunctional properties of MgO to design medicine-relevant devices, which exhibit both bioactive and antimicrobial properties.

3.8. Nanostructured V_2O_5

Recent studies have highlighted the ability of nanostructured V_2O_5 to mimic the myeloperoxidase activity [122, 123]. The activity is a characteristic of enzyme in human neutrophils, which eliminate bacteria via the catalysis of the hydrogen-peroxide-to-hypochlorite transformation in the presence of chloride ions [124]. This biomimetic property of V_2O_5 was effectively utilized for the processing of an anti-biofouling ship-hull coating using sea water as a source of hydrogen peroxide (100 nM) [123]. However, it has been shown that V_2O_5 generates ROS on its own [125], which indicated the possibility to perform a unique mode of antibacterial activity with a two-step mechanism: (i) generation of ROS and (ii) transformation of the generated ROS to antibacterially more potent hypochlorite ions. The use of V_2O_5 in medicine is limited by its relatively high solubility in aqueous media (>1 g/L). So-formed, high concentrations of vanadate ions are toxic to human cells [126, 127]. *In vitro* studies also showed their bi-phasic nature, as these ions stimulate proliferation of various types of mammalian cells at low concentrations (up to 10 μM) [128, 129]. They exhibit an insulin-mimicking action via the inhibition of tyrosine phosphatase [130]. Orally administered vanadates in rat models stimulated the orientation of the fibroblasts in parallel arrays early in the tissue-repair process, i.e., vanadate ions can accelerate tissue repair [131–133]. Vanadates improved the bone-formation rate, mechanical strength and mineralization [134], while the pro-oxidant potential of vanadates was not revealed in erythrocytes [135]. These studies confirmed the bioactive potential of vanadate ions when they are properly delivered, which might be effectively applied when designing the antibacterial drug-delivery system to enable controlled delivery of vanadate ions.

4. Concluding remarks

Antibacterial ions are prone to similar problems as antibiotics, i.e., biodistribution and bacterial resistance. Nevertheless, they offer new options, especially for local delivery, and the antibiotic resistant bacteria are not always resistant also to antibacterial ions, even though Cu- and Ag-resistance genes have been found associated with antibiotic resistance genes in a few cases. On the other hand, the major problem of nanoparticles is their non-selectivity and consequent toxicity for eukaryotic cells. For this reason, current findings are still far from a good substitution of antibiotics. It is very good that nanomaterials have many targets as opposed to antibiotics. This implies that they could be the solution for antibiotic resistance. But, the problem is that many of the targets are not specific for bacteria, in contrast to antibiotics. Particularly, the production of free radicals and reactive oxygen species in the absence of any cells needs to be avoided. Designing a wide therapeutic window (antibacterial activity at low concentrations and cytotoxicity at high concentrations of inorganic agent) is one of the greatest challenges for the application of inorganic antimicrobial agents. The possibility to modulate therapeutic window has the decision-making role in the perspective of inorganic antimicrobial agents as an alternative antimicrobial strategy.

Author details

Mario Kurtjak*, Nemanja Aničić and Marija Vukomanovicć

*Address all correspondence to: marija.vukomanovic@ijs.si

Advanced Materials Department, Jozef Stefan Institute, Ljubljana, Slovenia

References

[1] H. W. Boucher, G. H. Talbot, J. S. Bradley, J. E. Edwards, D. Gilbert, L. B. Rice, M. Scheld, B. Spellberg, and J. Bartlett. Bad bugs, no drugs: no ESKAPE! An update from the infectious diseases society of America. Clinical Infectious Diseases. 2009;**48**:1–12.

[2] H. J. Klasen. A historical review of the use of silver in the treatment of burns. I. Early uses. Burns. 2000;**26**:117–130.

[3] C. A. Moyer, L. Brentano, D. L. Gravens, H. W. Margraf, and W. W. Monafo. Treatment of large human burns with 0.5% silver nitrate solution. Archives Surgery. 1965;**90**:812–867.

[4] W. W. Monafo and C. A. Moyer. The treatment of extensive thermal burns with 0.5% silver nitrate solution. Annals of the New York Academy of Sciences. 1968;**150**:937–945.

[5] B. S. Atiyeh, M. Costagliola, S. N. Hayek, and S. A. Dibo. Effect of silver on burn wound infection control and healing: review of the literature. Burns. 2007;**33**:139–148.

[6] O. M. Goudouri, E. Kontonasaki, U. Lohbauer, and A. R. Boccaccini. Antibacterial properties of metal and metalloid ions in chronic periodontitis and peri-implantitis therapy. Acta Biomaterialia. 2014;**10**:3795–3810.

[7] S. Silver, L. T. Phung, and G. Silver. Silver as biocides in burn and wound dressings and bacterial resistance to silver compounds. Journal of Industrial Microbiology and Biotechnology. 2006;**33**:627–634.

[8] S. Chernousova and M. Epple. Silver as antibacterial agent: ion, nanoparticle, and metal. Angewandte Chemie – International Edition. 2013;**5**:1636–1653.

[9] K. Mijnendonckx, N. Leys, J. Mahillon, S. Silver, and R. Van Houdt. Antimicrobial silver: uses, toxicity and potential for resistance. Biometals. 2013;**26**:609–621.

[10] J. L. Hobman and L. C. Crossman. Bacterial antimicrobial metal ion resistance. Journal of Medical Microbiology. 2014;**64**:471–497.

[11] W. K. Jung, H. C. Koo, K. W. Kim, S. Shin, S. H. Kim, and Y. H. Park. Antibacterial activity and mechanism of action of the silver ion in *Staphylococcus aureus* and *Escherichia coli*. Applied and Environmental Microbiology. 2008;**74**:2171–2178.

[12] M. Rai, A. Yadav, and A. Gade. Silver nanoparticles as a new generation of antimicrobials. Biotechnology Advances. 2009;**27**:76–83.

[13] H. J. Park, J. Y. Kim, J. Kim, J. H. Lee, J. S. Hahn, M. B. Gu, and J. Yoon. Silver-ion-mediated reactive oxygen species generation affecting bactericidal activity. Water Research. 2009;**43**:1027–1032.

[14] G. Mulley, A. T. A. Jenkins, and N. R. Waterfield. Inactivation of the antibacterial and cytotoxic properties of silver ions by biologically relevant compounds. PLoS One. 2014;**9**:2–10.

[15] H. Dollwet and J. Sorenson. Historic uses of copper compounds in medicine. Trace Elements in Medicine. 1985;**2**:80–87.

[16] G. Borkow and J. Gabbay. Copper, an ancient remedy returning to fight microbial, fungal and viral infections. Current Chemical Biology. 2009;**3**:272–278.

[17] J. O'Gorman and H. Humphreys. Application of copper to prevent and control infection. Where are we now? Journal of Hospital Infection. 2012;**81**:217–223.

[18] J. Kolmas, E. Groszyk, and D. Kwiatkowska-Różycka. Substituted hydroxyapatites with antibacterial properties. BioMed Research International. 2014;**2014**:178–123.

[19] G. Grass, C. Rensing, and M. Solioz. Metallic copper as an antimicrobial surface. Applied and Environmental Microbiology. 2011;**77**:1541–1547.

[20] WHO. Trace elements in human nutrition and health. World Health Organization, http://www.who.int/nutrition/publications/micronutrients/9241561734/en/ ; 1996, p.p. 1-360.

[21] N. German, D. Doyscher, and C. Rensing. Bacterial killing in macrophages and amoeba: do they all use a brass dagger? Future Microbiology. 2013;**8**:1257–1264.

[22] M. Valko, H. Morris, and M. T. D. Cronin. Metals, toxicity and oxidative stress. Current Medicinal Chemistry. 2005;**12**:1161–1208.

[23] J. A. Lemire, J. J. Harrison, and R. J. Turner. Antimicrobial activity of metals: mechanisms, molecular targets and applications. Nature Reviews Microbiology. 2013;**11**:371–384.

[24] R. B. Thurman and C. P. Gerba. The molecular mechanisms of copper and silver ion disinfection of bacteria and viruses. Critical Reviews in Environmental Control. 1989;**18**: 295–315.

[25] G. Borkow and J. Gabbay. Copper as a biocidal tool. Current Medicinal Chemistry. 2005;**12**:2163–2175.

[26] A. B. G. Landsdown. Zinc in the healing wound. Lancet. 1996;**347**:706–707.

[27] J. C. King, D. M. Shames, and L. R. Woodhouse. Zinc and health: current status and future directions zinc homeostasis in humans 1. Journal of Nutrition. 2000;**130**:1360–1366.

[28] M. Jakubowski. Zinc and cadmium compounds. In: E. Bingham and B. Cohrssen, editors. Patty's toxicology. John Wiley & Sons ed; 2012, p.p. 167-212. DOI: 10.1002/0471435139.tox029.pub2.

[29] E. Coleman. Zinc proteins: enzymes, storage proteins, transcription factors, and replication proteins. Annual Review of Biochemistry. 1992;**61**:897–946.

[30] T. N. Phan, T. Buckner, J. Sheng, J. D. Baldeck, and R. E. Marquis. Physiologic actions of zinc related to inhibition of acid and alkali production by oral streptococci in suspensions and biofilms. Oral Microbiology and Immunology. 2014;**19**:31–38.

[31] B. Aydin Sevinç and L. Hanley. Antibacterial activity of dental composites containing zinc oxide nanoparticles. Journal of Biomedical Materials Research Part B: Applied Biomaterials. 2010;**94B**:22–31.

[32] J. T. Seil and T. J. Webster. Antimicrobial applications of nanotechnology: methods and literature. International Journal of Nanomedicine. 2012;**7**:2767–2781.

[33] J. Sheng, P. T. M. Nguyen, and R. E. Marquis. Multi-target antimicrobial actions of zinc against oral anaerobes. Archives of Oral Biology. 2005;**50**:747–757.

[34] D. Cummins. Zinc citrate/triclosan: a new anti-plaque system for the control of plaque and the prevention of gingivitis: short-term clinical and mode of action studies. Journal of Clinical Periodontology. 1991;**18**:455–461.

[35] R. Choudhury and S. Srivastava Zinc resistance mechanisms in bacteria. Current Science. 2001;**81**:768–775.

[36] L. M. Cavaco, H. Hasman, and F. M. Aarestrup. Zinc resistance of *Staphylococcus aureus* of animal origin is strongly associated with methicillin resistance. Veterinary Microbiology. 2011;**150**:344–348.

[37] C. Bonchi, F. Imperi, F. Minandri, P. Visca, and E. Frangipani. Repurposing of gallium-based drugs for antibacterial therapy. Biofactors. 2014;**40**:303–312.

[38] A. Rangel-Vega, L. R. Bernstein, E. A. Mandujano-Tinoco, S. J. García-Contreras, and R. García-Contreras. Drug repurposing as an alternative for the treatment of recalcitrant bacterial infections. Frontiers in Microbiology. 2015;**6**:1–8.

[39] L. R. Bernstein. Mechanisms of therapeutic activity for gallium. Pharmacological Reviews. 1998;**50**:665–682.

[40] O. Olakanmi, B. E. Britigan, and L. S. Schlesinger. Gallium disrupts iron metabolism of mycobacteria residing within human macrophages. Infection and Immunity. 2000;**68**:5619–5627.

[41] Y. Kaneko, M. Thoendel, O. Olakanmi, B. E. Britigan, and P. K. Singh. The transition metal gallium disrupts *Pseudomonas aeruginosa* iron metabolism and has antimicrobial and antibiofilm activity. The Journal of Clinical Investigation. 2007;**117**:877–888.

[42] K. DeLeon, F. Balldin, C. Watters, A. Hamood, J. Griswold, S. Sreedharan, and P. Rumbaugh. Gallium maltolate treatment eradicates *Pseudomonas aeruginosa* infection in thermally injured mice. Antimicrobial Agents and Chemotherapy. 2009;**53**:1331–1337.

[43] O. Olakanmi, B. Kesavalu, R. Pasula, M. Y. Abdalla, L. S. Schlesinger, and B. E. Britigan. Gallium nitrate is efficacious in murine models of tuberculosis and inhibits key bacterial Fe-dependent enzymes. Antimicrobial Agents and Chemotherapy. 2013;**57**:6074–6080.

[44] C. H. Goss, D. B. Hornick, M. L. Aitken, G. Anderson, and E. Caldwell. Phase 1 pharmacokinetic and safety study of intravenous ganite (gallium nitrate) in cf. Pediatric Pulmonology. 2012;**47**:303.

[45] E. Banin, A. Lozinski, K. M. Brady, E. Berenshtein, P. W. Butterfield, M. Moshe, M. Chevion, E. P. Greenberg, and E. Banin. The potential of desferrioxamine-gallium as an anti-Pseudomonas therapeutic agent. Proceedings of the National Academy of Sciences of the United States of America. 2008;**105**:16761–16766.

[46] O. Rzhepishevska, B. Ekstrand-Hammarström, M. Popp, E. Björn, A. Bucht, A. Sjöstedt, H. Antti, and M. Ramstedt. The antibacterial activity of Ga3+ is influenced by ligand complexation as well as the bacterial carbon source. Antimicrobial Agents and Chemotherapy. 2011;**55**:5568–5580.

[47] I. Stojiljkovic, V. Kumar, and N. Srinivasan. Non-iron metalloporphyrins: potent antibacterial compounds that exploit haem/Hb uptake systems of pathogenic bacteria. Molecular Microbiology. 1999;31:429-42.

[48] A. B. Kelson, M. Carnevali, and V. Truong-Le. Gallium-based anti-infectives: targeting microbial iron-uptake mechanisms. Current Opinion in Pharmacology. 2013;**13**:707–716.

[49] L. R. Bernstein. Gallium, therapeutic effects. In: R. H. Kretsinger, V. N. Uversky, and E. A. Permyakov, editors. Encyclopedia of metalloproteins. New York: Springer; 2013, pp. 823–835.

[50] C. R. Chitambar. Medical applications and toxicities of gallium compounds. International Journal of Environmental Research and Public Health. 2010;**7**:2337–2361.

[51] R. Bériault, R. Hamel, D. Chenier, R. J. Mailloux, H. Joly, and V. D. Appanna. The overexpression of NADPH-producing enzymes counters the oxidative stress evoked by gallium, an iron mimetic. Biometals. 2007;**20**:165–176.

[52] S. P. Valappil, D. Ready, E. A. Abou Neel, D. M. Pickup, L. A. O'Dell, W. Chrzanowski, J. Pratten, R. J. Newport, M. E. Smith, M. Wilson, and J. C. Knowles. Controlled delivery of antimicrobial gallium ions from phosphate-based glasses. Acta Biomaterialia. 2009;**5**:1198–1210.

[53] S. Pourshahrestani, E. Zeimaran, N. Adib Kadri, N. Gargiulo, S. Samuel, S. V. Naveen, T. Kamarul, and M. R. Towler. Gallium-containing mesoporous bioactive glass with potent hemostatic activity and antibacterial efficacy. Journal of Materials Chemistry B. 2016;**4**:71–86.

[54] R. Sahdev, T. I. Ansari, S. M. Higham, and S. P. Valappil. Potential use of gallium-doped phosphate-based glass material for periodontitis treatment. Journal of Biomaterials Applications. 2015;**30**:85–92.

[55] S. P. Valappil, H. H. P. Yiu, L. Bouffier, C. K. Hope, G. Evans, J. B. Claridge, S. M. Higham, and M. J. Rosseinsky. Effect of novel antibacterial gallium-carboxymethyl cellulose on *Pseudomonas aeruginosa*. Dalton Transactions. 2013;**42**:1778–1786.

[56] S. Shruti, A. J. Salinas, G. Lusvardi, G. Malavasi, L. Menabue, and M. Vallet-Regi. Mesoporous bioactive scaffolds prepared with cerium-, gallium- and zinc-containing glasses. Acta Biomaterialia. 2013;**9**:4836–4844.

[57] M. Franchini, G. Lusvardi, G. Malavasi, and L. Menabue. Gallium-containing phosphosilicate glasses: synthesis and in vitro bioactivity. Materials Science and Engineering C. 2012;**32**:1401–1406.

[58] G. Lusvardi, G. Malavasi, L. Menabue, and S. Shruti. Gallium-containing phosilicate glasses: functionalization and in-vitro bioactivity. Materials Science and Engineering C. 2013;**33**:3190–3196.

[59] A. Cochis, B. Azzimonti, C. Della Valle, R. Chiesa, C. R. Arciola, and L. Rimondini. Biofilm formation on titanium implants counteracted by grafting gallium and silver ions. Journal of Biomedical Materials Research – A. 2015;**103**:1176–1187.

[60] E. Frangipani, C. Bonchi, F. Minandri, F. Imperi, and P. Visca. Pyochelin potentiates the inhibitory activity of gallium on *Pseudomonas aeruginosa*. Antimicrobial Agents and Chemotherapy. 2014;**58**:5572–5575.

[61] R. García-Contreras, E. Lira-Silva, R. Jasso-Chávez, I. L. Hernández-González, T. Maeda, T. Hashimoto, F. C. Boogerd, L. Sheng, T. K. Wood, and R. Moreno- Sánchez. Isolation and characterization of gallium resistant *Pseudomonas aeruginosa* mutants. International Journal of Medical Microbiology: IJMM. 2013;**303**:574–582.

[62] E.-J. Yang, J. Jang, S. Kim, and I.-H. Choi. Silver nanoparticles as a smart antimicrobial agent. Journal of Bacteriology and Virology. 2012;**42**:177.

[63] S. Pal, Y. K. Tak, and J. M. Song. Does the antibacterial activity of silver nanoparticles depend on the shape of the nanoparticle? A study of the gram-negative bacterium *Escherichia coli*. Applied and Environmental Microbiology. 2007;**73**:1712–1720.

[64] H. Zhang and G. Chen. Potent antibacterial activities of Ag/TiO_2 nanocomposite powders synthesized by a one-pot sol-gel method. Environmental Science & Technology. 2009;**43**:2905–2910.

[65] J. R. Morones, J. L. Elechiguerra, A. Camacho, K. Holt, J. B. Kouri, J. T. Ramírez, and M. J. Yacaman. The bactericidal effect of silver nanoparticles. Nanotechnology. 2005;**16**: 2346–2353.

[66] S. H. Kim, H. S. Lee, D. S. Ryu, S. J. Choi, and D. S. Lee. Antibacterial activity of silver-nanoparticles against *Staphylococcus aureus* and *Escherichia coli*. Korean Journal of Microbiology and Biotechnology. 2011;**39**:77–85.

[67] W. He, Y. T. Zhou, W. G. Wamer, M. D. Boudreau, and J. J. Yin. Mechanisms of the pH dependent generation of hydroxyl radicals and oxygen induced by Ag nanoparticles. Biomaterials. 2012;**33**:7547–7555.

[68] R. P. Singh and P. Ramarao. Cellular uptake, intracellular trafficking and cytotoxicity of silver nanoparticles. Toxicology Letters. 2012;**213**:249–259.

[69] Z.-M. Xiu, Q.-B. Zhang, H. L. Puppala, V. L. Colvin, and P. J. J. Alvarez. Negligible particle-specific antibacterial activity of silver nanoparticles. Nano Letters. 2012;**12**:4271–4275.

[70] R. Foldbjerg, X. Jiang, T. Miclăuş, C. Chen, H. Autrup, and C. Beer. Silver nanoparticles – wolves in sheep's clothing? Toxicology Research. 2015; 4: 563-75.

[71] B. Reidy, A. Haase, A. Luch, K. A. Dawson, and I. Lynch. Mechanisms of silver nanoparticle release, transformation and toxicity: a critical review of current knowledge and recommendations for future studies and applications. Materials. 2013;**6**:2295–2350.

[72] N. Durán, M. Durán, M. B. de Jesus, A. B. Seabra, W. J. Fávaro, and G. Nakazato. Silver nanoparticles: a new view on mechanistic aspects on antimicrobial activity. Nanomedicine: Nanotechnology, Biology, and Medicine. 2016;**12**:789–799.

[73] S. J. Yu, J. B. Chao, J. Sun, Y. G. Yin, J. F. Liu, and G. B. Jiang. Quantification of the uptake of silver nanoparticles and ions to HepG2 cells. Environmental Science and Technology. 2013;**47**:3268–3274.

[74] X. Jiang, T. Miclăus, L. Wang, R. Foldbjerg, D. S. Sutherland, H. Autrup, C. Chen, and C. Beer. Fast intracellular dissolution and persistent cellular uptake of silver nanoparticles in CHO-K1 cells: implication for cytotoxicity. Nanotoxicology. 2015;**9**:1743–5390.

[75] S. Agnihotri, S. Mukherji, and S. Mukherji. Immobilized silver nanoparticles enhance contact killing and show highest efficacy: elucidation of the mechanism of bactericidal action of silver. Nanoscale. 2013;**5**:7328–7340.

[76] J. L. Graves, M. Tajkarimi, Q. Cunningham, A. Campbell, H. Nonga, S. H. Harrison, and J. E. Barrick. Rapid evolution of silver nanoparticle resistance in Escherichia coli. Frontiers in Genetics. 2015;6(42):1–13.

[77] V. R. Rai and J. A. Bai. Nanoparticles and their potential application as antimicrobials. In: A. Méndez-Vilas, editor. Science against microbial pathogens: communicating current research and technological advances. Formatex; 2011, pp. 197–209.

[78] G. Applerot, J. Lellouche, A. Lipovsky, Y. Nitzan, R. Lubart, A. Gedanken, and E. Banin. Understanding the antibacterial mechanism of CuO nanoparticles: revealing the route of induced oxidative stress. Small. 2012;**8**:3326–3337.

[79] H. L. Karlsson, P. Cronholm, J. Gustafsson, and L. Möller. Copper oxide nanoparticles are highly toxic: a comparison between metal oxide nanoparticles and carbon nanotubes. Chemical Research in Toxicology. 2008;**21**:1726–1732.

[80] Y. W. Baek and Y. J. An. Microbial toxicity of metal oxide nanoparticles (CuO, NiO, ZnO, and Sb_2O_3) to *Escherichia coli, Bacillus subtilis,* and *Streptococcus aureus*. Science of the Total Environment. 2011;**409**:1603–1608.

[81] M. Horie, K. Fujita, H. Kato, S. Endoh, K. Nishio, L. K. Komaba, et al. Association of the physical and chemical properties and the cytotoxicity of metal oxide nanoparticles: metal ion release, adsorption ability and specific surface area. Metallomics. 2012;**4**:350–360.

[82] A. Sirelkhatim, S. Mahmud, A. Seeni, N. H. M. Kaus, L. C. Ann, S. K. M. Bakhori, H. Hasan, and D. Mohamad. Review on zinc oxide nanoparticles: antibacterial activity and toxicity mechanism. Nanomicro Letters. 2015;**7**:219–242.

[83] D. Zvekić, V. V. Srdić, M. A. Karaman, and M. N. Matavulj. Antimicrobial properties of ZnO nanoparticles incorporated in polyurethane varnish. Processing and Application of Ceramics. 2011;**5**:41–45.

[84] Y. H. Leung, X. Xu, A. P. Y. Ma, F. Liu, A. M. C. Ng, Z. Shen, et al. Toxicity of ZnO and TiO_2 to *Escherichia coli* cells. Scientific Reports. 2016;**6**:35243.

[85] A. J. Huh and Y. J. Kwon. Nanoantibiotics': a new paradigm for treating infectious diseases using nanomaterials in the antibiotics resistant era. Journal of Controlled Release. 2011;**156**:128–145.

[86] R. Dastjerdi and M. Montazer. A review on the application of inorganic nano-structured materials in the modification of textiles: focus on anti-microbial properties. Colloids and Surfaces B: Biointerfaces. 2010;**79**:5–18.

[87] A. M. Alkilany and C. J. Murphy. Toxicity and cellular uptake of gold nanoparticles: what we have learned so far? Journal of Nanoparticle Research. 2010;**12**:2313–2333.

[88] C. Lasagna-Reeves, D. Gonzalez-Romero, M. A. Barria, I. Olmedo, A. Clos, V. M. Sadagopa Ramanujam, A. Urayama, L. Vergara, M. J. Kogan, and C. Soto. Bioaccumulation and toxicity of gold nanoparticles after repeated administration in mice. Biochemical and Biophysical Research Communications. 2010;**393**:649–655.

[89] R. Shukla, V. Bansal, M. Chaudhary, A. Basu, R. R. Bhonde, and M. Sastry. Bio-compatibility of gold nanoparticles and their endocytotic fate inside the cellular compartment: a microscopic overview. Langmuir. 2005;**21**:10644–10654.

[90] N. Khlebtsov and L. Dykman. Biodistribution and toxicity of engineered gold nanoparticles: a review of in vitro and in vivo studies. Chemical Society Reviews. 2011;**40**: 1647–1671.

[91] L. Shang, K. Nienhaus, and G. U. Nienhaus. Engineered nanoparticles interacting with cells: size matters. Journal of Nanobiotechnology. 2014;**12**:5.

[92] B. D. Chithrani, A. A. Ghazani, and W. C. W. Chan. Determining the size and shape dependence of gold nanoparticle uptake into mammalian cells. Nano Letters. 2006;**6**: 662–668.

[93] D. Yohan, C. Cruje, X. Lu, and D. B. Chithrani. Size-dependent gold nanoparticle interaction at nanomicro interface using both monolayer and multilayer (tissue- like) cell models. Nanomicro Letters. 2016;**8**:44–53.

[94] Y.-H. Gao, N.-C. Zhang, Y.-W. Zhong, H.-H. Cai, and Y.-l. Liu. Preparation and characterization of antibacterial Au/C core-shell composite. Applied Surface Science. 2010;**256**: 6580–6585.

[95] Y. Zhao, Y. Tian, Y. Cui, W. Liu, W. Ma, and X. Jiang. Small molecule-capped gold nanoparticles as potent antibacterial agents that target gram-negative bacteria. Journal of the American Chemical Society. 2010;**132**:12349–12356.

[96] M. Vukomanović, S. D. Škapin, and D. Suvorov. Functionalized hydroxyapatite/gold composites as "green" materials with antibacterial activity and a process for preparing and use thereof. 2013. WO2013187846 (A1), PCT/SI2013/000025, P-201200204, BIL 2013/12,31.12.2013.

[97] M. Vukomanović, M. Logar, S. D. Škapin, and D. Suvorov. Hydroxyapatite/gold/arginine: designing the structure to create antibacterial activity. Journal of Materials Chemistry B. 2014;**2014**:1557–1564.

[98] Y. Cui, Y. Zhao, Y. Tian, W. Zhang, X. Lü, and X. Jiang. The molecular mechanism of action of bactericidal gold nanoparticles on *Escherichia coli*. Biomaterials. 2012;**33**:2327–2333.

[99] P. S. Murthy, V. P. Venugopalan, P. Sahoo, S. Dhara, A. Das, and A. K. Tyagi. Gallium oxide nanoparticle induced inhibition of bacterial adhesion and biofilm formation. In: 2011 International Conference on Nanoscience, Engineering and Technology (Iconset); 2011, pp. 490–493.

[100] K. Girija, S. Thirumalairajan, and D. Mangalaraj. Morphology controllable synthesis of parallely arranged single-crystalline Ga_2O_3 nanorods for photocatalytic and antimicrobial activities. Chemical Engineering Journal. 2014;**236**:181–190.

[101] Z. Tzitrinovich, A. Lipovsky, A. Gedanken, and R. Lubart. Visible light-induced OH radicals in Ga_2O_3: an EPR study. Physical Chemistry Chemical Physics. 2013;**15**:12977–12981.

[102] P. Sahoo, P. S. Murthy, S. Dhara, V. P. Venugopalan, A. Das, and A. K. Tyagi. Probing the cellular damage in bacteria induced by GaN nanoparticles using confocal laser Raman spectroscopy. Journal of Nanoparticle Research. 2013;**15**:1841.

[103] M. S. Kandanapitiye, B. Valley, L. D. Yang, A. M. Fry, P. M. Woodward, and S. D. Huang. Gallium analogue of soluble Prussian blue KGa[Fe(CN)6]·nH_2O: synthesis, characterization, and potential biomedical applications. Inorganic Chemistry. 2013;**52**:2790–2792.

[104] Y. Lu, Q. Hu, Y. Lin, D. B. Pacardo, C. Wang, W. Sun, F. S. Ligler, M. D. Dickey, and Z. Gu. Transformable liquid-metal nanomedicine. Nature Communications. 2015;**6**:10066.

[105] P. Narayanasamy, B. L. Switzer, and B. E. Britigan. Prolonged-acting, Multi-targeting gallium nanoparticles potently inhibit growth of both HIV and mycobacteria in co-infected human macrophages. Scientific Reports. 2015;**5**:8825.

[106] M. Managa, E. Antunes, and T. Nyokong. Conjugates of platinum nanoparticles with gallium tetra-(4-carboxyphenyl) porphyrin and their use in photodynamic antimicrobial chemotherapy when in solution or embedded in electrospun fiber. Polyhedron. 2014;**76**:94–101.

[107] M. Kurtjak, M. Vukomanović, L. Kramer, and D. Suvorov. Biocompatible nano-gallium/hydroxyapatite nanocomposite with antimicrobial activity. Journal of Materials Science: Materials in Medicine. 2016;**27**:170.

[108] J. Sawai and T. Yoshikawa. Quantitative evaluation of antifungal activity of metallic oxide powders (MgO, CaO and ZnO) by an indirect conductimetric assay. Journal of Applied Microbiology. 2004;**96**:803–809.

[109] J. Sawai, H. Kojima, H. Igarashi, A. Hashimoto, S. Shoji, T. Sawaki, et al. Antibacterial characteristics of magnesium oxide powder. World Journal of Microbiology and Biotechnology. 2000;**16**:187–194.

[110] K. Krishnamoorthy, G. Manivannan, S. J. Kim, K. Jeyasubramanian, and M. Premanathan. Antibacterial activity of MgO nanoparticles based on lipid peroxidation by oxygen vacancy. Journal of Nanoparticle Research. 2012;**14**:1063–1072.

[111] Y. He, S. Ingudam, S. Reed, A. Gehring, T. P. Strobaugh, and P. Irwin. Study on the mechanism of antibacterial action of magnesium oxide nanoparticles against food-borne pathogens. Journal of Nanobiotechnology. 2016;**14**:54.

[112] Y. H. Leung, A. M. C. Ng, X. Xu, Z. Shen, L. A. Gethings, M. T. Wong, et al. Mechanisms of antibacterial activity of MgO: non-ROS mediated toxicity of MgO nanoparticles towards *Escherichia coli*. Small. 2014;**10**:1171–1183.

[113] F. Guilhelmelli, N. Vilela, P. Albuquerque, L. da S. Derengowski, I. Silva-Pereira, and C. M. Kyaw. Antibiotic development challenges: the various mechanisms of action of antimicrobial peptides and of bacterial resistance. Frontiers in Microbiology. 2013;**4**:1–12.

[114] M. Anpo, M. Che, B. Fubini, and E. Garrone. Generation of superoxide ions at oxide surfaces. Topics in Catalysis. 1999;**8**:189–198.

[115] P. K. Stoimenov, R. L. Klinger, G. L. Marchin, and K. J. Klabunde. Metal oxide nanoparticles as bactericidal agents. Langmuir. 2002;**18**:6679–6686.

[116] L. Huang, D. Q. Li, Y. J. Lin, M. Wei, D. G. Evans, and X. Duan. Controllable preparation of Nano-MgO and investigation of its bactericidal properties. Journal of Inorganic Biochemistry. 2005;**99**:986–993.

[117] Y. Rao, W. Wang, F. Tan, Y. Cai, J. Lu, and X. Qiao. Sol-gel preparation and antibacterial properties of Li-doped MgO nanoplates. Ceramics International. 2014;**40**:14397–14403.

[118] A. Balakrishnan, P. Pizette, C. L. Martin, S. V. Joshi, and B. P. Saha. Effect of particle size in aggregated and agglomerated ceramic powders. Acta Materialia. 2010;**58**:802–812.

[119] N. Anicic, M. Vukomanovic, and D. Suvorov. The nano-texturing of MgO microrods for antibacterial applications. RSC Advances. 2016;**6**:102657–102664.

[120] R. Swaminathan. Disorders of magnesium metabolism. Clinical Biochemistry Review. 2003;**24**:47–66.

[121] T. Motoyama, H. Sano, and H. Fukuzaki. Oral magnesium supplementation in patients with essential hypertension. Hypertension. 1989;**13**:227–232.

[122] K. Krishnamoorthy, J. Y. Moon, H. B. Hyun, S. K. Cho, and S.-J. Kim. Mechanistic investigation on the toxicity of MgO nanoparticles toward cancer cells. Journal of Materials Chemistry. 2012;**22**:24610–24617.

[123] R. André, F. Natálio, M. Humanes, J. Leppin, K. Heinze, R. Wever, et al. V_2O_5 nanowires with an intrinsic peroxidase-like activity. Advanced Functional Materials. 2011;**21**: 501–509.

[124] F. Natalio, R. André, A. F. Hartog, B. Stoll, K. P. Jochum, R. Wever, et al. Vanadium pentoxide nanoparticles mimic vanadium haloperoxidases and thwart biofilm formation. Nature Nanotechnology. 2012;**7**:530–535.

[125] B. Amulic, C. Cazalet, G. L. Hayes, K. D. Metzler, and A. Zychlinsky. Neutrophil function: from mechanisms to disease. Annual Review of Immunology. 2012;**30**:459–489.

[126] A. Galanis, A. Karapetsas, and R. Sandaltzopoulos. Metal-induced carcinogenesis, oxidative stress and hypoxia signalling. Mutation Research-Genetic Toxicology and Environmental Mutagenesis. 2009;**674**:31–35.

[127] S. Ivanković, S. Musić, M. Gotić, and N. Ljubesić. Cytotoxicity of nanosize V_2O_5 particles to selected fibroblast and tumor cells. Toxicology. In Vitro. 2006;**20**:286–294.

[128] J. D. Jarrell, B. Dolly, and J. R. Morgan. Controlled release of vanadium from titanium oxide coatings for improved integration of soft tissue implants. Journal of Biomedical Materials Research Part A. 2009;**90**:272–281.

[129] D. A. Barrio, M. D. Braziunas, S. B. Etcheverry, and A. M. Cortizo. Maltol complexes of vanadium (IV) and (V) regulate in vitro alkaline phosphatase activity and osteoblast-like cell growth. Journal of Trace Elements in Medicine and Biology. 1997;**11**:110–115.

[130] A. M. Cortizo, V. C. Sálice, C. M. Vescina, and S. B. Etcheverry. Proliferative and morphological changes induced by vanadium compounds on Swiss 3T3 fibroblasts. Biometals. 1997;**10**:127–133.

[131] K. E. Moyer, A. A. Saba, R. M. Hauck, and H. P. Ehrlich. Systemic vanadate ingestion modulates rat tendon repair. Experimental and Molecular Pathology. 2003;**75**:80–88.

[132] H. P. Ehrlich, K. A. Keefer, G. O. Maish, R. L. Myers, and D. R. Mackay. Vanadate ingestion increases the gain in wound breaking strength and leads to better organized collagen fibers in rats during healing. Plastic and Reconstructive Surgery. 2001;**107**:471–477.

[133] M. Y. Lee and H. P. Ehrlich. Influence of vanadate on migrating fibroblast orientation within a fibrin matrix. Journal of Cellular Physiology. 2008;**217**:72–76.

[134] D. M. Facchini, V. G. Yuen, M. L. Battell, J. H. McNeill, and M. D. Grynpas. The effects of vanadium treatment on bone in diabetic and non-diabetic rats. Bone. 2006;**38**:368–377.

[135] A. Ścibior, H. Zaporowska, A. Wolińska, and J. Ostrowski. Antioxidant enzyme activity and lipid peroxidation in the blood of rats co-treated with vanadium (V+5) and chromium (Cr+3). Cell Biology and Toxicology. 2010;**26**:509–526.

Chapter 4

Current Approaches for Exploration of Nanoparticles as Antibacterial Agents

Didem Şen Karaman, Suvi Manner,
Adyary Fallarero and Jessica M. Rosenholm

Additional information is available at the end of the chapter

http://dx.doi.org/10.5772/68138

Abstract

The ascending anxiety regarding antimicrobial resistance as well as the recalcitrant nature of biofilm-associated infections call for the development of alternative strategies to treat bacterial diseases. Nanoparticles have been considered as one of the emerging and promising platforms in this respect. Their unique physical and chemical properties may lead to fine-tuned interactions between them and bacteria. In this chapter, we aim to provide an overview on the use of nanoparticles as antimicrobial agents. Both antibacterial and anti-biofilm activities of nanoparticles and their current approaches will be reviewed. The *in vitro* methods that are used to evaluate the potency of nanoparticles as antimicrobial agents will be discussed in detail.

Keywords: antibacterial agents, nanoparticles, antibacterial resistance (AMR), anti-biofilm agents, *in vitro* methods

1. Introduction

The term antimicrobial was derived from the Greek words anti (against), micro (little), and bios (life), and it refers to all agents that act against microorganisms. Thus, antimicrobials include agents that act against bacteria (antibacterial), viruses (antiviral), fungi (antifungal), and protozoa (antiprotozoal). Among these, antibacterial agents are by far the most widely known and studied class of antimicrobials. Nowadays, the emergence of antimicrobial resistance (AMR) among the microbial pathogens greatly increases the threat generated by bacterial infections. Drug-resistant bacteria lead to poor clinical outcomes increasing health care costs and mortality. In the US, the estimated health care costs associated with the treatment of infectious diseases are annually more than 120 billion dollars, and further, treatment of infections caused by resistant pathogens costs 5 billion dollars per year [1]. According to the US Center for Disease Control and Prevention (CDC), more than two million antibiotic-resistant infections occur every year in the US and they lead to 23,000 deaths [2]. In the European Union, antibiotic-resistant infections are responsible for 25,000 deaths every year [3]. Both Gram-positive, especially methicillin-resistant *Staphylococcus aureus* (MRSA), vancomycin-resistant *Enterococcus faecium* (VRE), and drug-resistant *Streptococcus pneumoniae*, and Gram-negative bacteria, namely multidrug-resistant *Acinetobacter baumannii* (MRAB), carbapenem-resistant *Enterobacteriaceae* (CRE), and *Pseudomonas aeruginosa*, display resistance to multiple drugs and are of serious concern [4]. In addition, biofilm formation complicates treatment of various infections. Biofilm-related infections, such as chronic wounds and urinary tract infections, pneumonia in cystic fibrosis patients, and infections related to the use of medical devices, comprise up to 80% of all human bacterial infections, and affect millions of people resulting in up to 550,000 deaths every year [5–7]. The emergence of AMR has been mainly attributed to the inappropriate and excessive use of antimicrobials in humans and animals. In many countries, unregulated availability of antibiotics without prescription results in promotion of overuse [8]. Further, inappropriate prescription of antibiotics also contributes to the promotion of resistant bacteria. The reports have shown that the diagnosis, choice of treatment agent, or duration of antibiotic therapy were incorrect in 30–50% of the cases [9], and 30–60% of antibiotics prescribed in intensive care units have been inappropriate or unnecessary [10]. Extensive agricultural usage of antibiotics in livestock as growth supplement is another reason of excessive consumption of antibiotics by humans, through the intake of resistant bacteria in the food supplies.

It is estimated that more than 70% of all pathogenic bacteria are resistant to at least one of the conventional antibiotics [11]. Antimicrobial resistance is acquired on both cellular and community levels [12]. Acquirement and dissemination of resistance genes is a process that occurs over time. Nevertheless, the evolution of bacterial resistance is substantially accelerated by the dispensable use of antibiotics [13]. Further dissemination of resistance genes between bacterial species has led to the emergence of multidrug-resistant (MDR) bacteria [14]. Community level of resistance is caused by biofilm formation [15]. However, when it comes to biofilms, the genetically transferable, conventional resistance mechanisms are not the leading cause of decreased antimicrobial susceptibility [7]. Bacterial biofilms are structured communities of bacteria embedded in a matrix of extracellular polymeric substances (EPS) that can be formed on variety of surfaces, such as tissues and medical devices [16]. Biofilm is a transient phenotype that makes even sensitive bacteria without known genetic basis for resistance to display remarkably reduced susceptibility to antimicrobials and host immune responses [17]. Many factors contribute to the antimicrobial tolerance of biofilms. First, biofilm matrix can restrict penetration of antibiotics and protects the cells from detrimental insults [18]. Secondly, biofilms comprise a heterogeneous population of cells that are in different physiological states due to decreasing oxygen and nutrient gradients existing between the surface and deeper layers of biofilms. For example, cells located in the deepest part of the biofilm tend to display a slower growth rate and, therefore, are less susceptible to antibiotics that are developed against dividing cells [19, 20]. Further, the non-dividing, dormant population of bacteria, referred also as persister cells, is in well-protected mode and highly tolerant antibiotics. These cells survive even from prolonged antibiotic treatment and serve as reservoirs for infections [21]. Consequently, biofilm bacteria can be up to 1000 times more tolerant to antimicrobial agents than planktonic cells of the same species [22].

Despite this, pharmaceutical companies have substantially declined investments in antimicrobial drug discovery during the past few decades [23]. Antimicrobial drug discovery is not economically attractive, and regulatory requirements have become very challenging [3, 24]. The need of novel bactericidal agents has increased due to the emergence of multi-drug resistant bacterial strains and biofilm-associated infections. Consequently, attention has been especially devoted to emerging nanoparticle-based materials in the field of antimicrobial therapies. In this chapter, the existing nanoparticles as antimicrobial means and the current *in vitro* test methods that will ease clinical translation of nanomaterials by establishing *in vivo* relevant data will be described and discussed.

2. Nanotechnology-based antibacterial therapies

Antibacterial applications of nanotechnology are gaining importance to prevent the catastrophic consequences of antibiotic resistance. Nanotechnology can be implemented as preventives, diagnostics, drug carriers, and synergetics in the antibacterial therapies.

The unique properties of nanomaterials compared to its bulk form make them favourable for antibacterial therapies. Many inorganic and organic nanomaterials represent inherent antibacterial properties that are not expressed in their bulk form. Fast and sensitive bacterial detection can be provided with nanoparticle-based approaches. Furthermore, nanoparticles offer discrete advantages as antibacterial drug delivery systems. They can be designed as targeted, environmentally responsive, combinatorial delivery systems [25]. Another approach of nanomaterials for the antibacterial therapy is as vaccine that contains nanoparticles as adjuvants or delivery vehicles, which provoke immune responses against bacterial infection. In the following parts in section 2.1 and 2.2, the existing nanotechnologies for the antibacterial delivery systems and inherently antibacterial nanoparticles will be discussed in detail.

2.1. Nanomaterials as antibacterial delivery systems

The existing disadvantages of conventional antibiotics can be solved to some extent by using nanomaterial-based antimicrobial delivery systems. In such approaches, the conventional antibiotics can be loaded into the nanoparticles through physical encapsulation, adsorption, or chemical conjugation. By this way, the pharmacokinetics and therapeutic index of the drug can ideally be improved compared to the free form of the drug. The aimed-for advantages are provided by the improved serum solubility, prolonged systemic circulation lifetime of the drug, targeted delivery of the drug to the site of infection, sustained and controlled release of the drug, and also combinatorial drug delivery to the site of interest that could be reached by virtue of the nanoscopic delivery system [26–28]. This rationale of nanotherapeutics in this case aimed to enhance the therapeutic effect and minimize the side effects of antibiotics, starts with the appropriate design of nanoparticles. In nanoparticles design, the particle size, surface properties, and the release profile of the therapeutic agent have vital impact on the success of the therapeutic approach. Various nanoparticles-based drug delivery systems have been designed and investigated for improving the efficacy of antibiotics of the administered drugs, the most common of which shall be outlined in the following.

Lipid-based nanoparticles are widely used for the delivery of antibacterial agents. They can be designed as liposomes, solid lipid nanoparticles (SLN), and nanostructured lipid carriers (NLC). Liposomes are one of the most studied nanosystems for antimicrobial therapy in various diseases. Liposomes are spherical lipid vesicles with bilayered membrane structure, consisting of amphiphilic lipid molecules. Since their structure is similar to the bacterial cell membrane, efficacious interaction between liposomes and cells can be obtained. These interactions may create adsorption, endocytosis, lipid exchange, and fusion of the liposomes. Especially, the design of liposomes that cause fusion and is known as fusogenic is the most attractive one in the sense of efficiency. Fusogenic liposomes are able to destabilize the bacterial membrane and release their therapeutic content inside the cells [29, 30]. The structure of liposomes, where an aqueous cavity is surrounded by lipid membranes, empowers them to transport both lipophilic and hydrophilic drugs (in lipid bilayers and aqueous compartments, respectively) without chemical modification, protecting them from degradation [31]. SLNs are composed of a solid lipid core stabilized by surfactants and are moderately amorphous structures in which bilayers are not distinguished. They can provide long-term stability and better incorporation efficacy for hydrophobic drugs and can be easily scaled-up in production. NLCs were developed in order to overcome the limitations of SLNs regarding low-loading capacity for nonhydrophobic drugs and their stability issues. In the NLC structure, liquid lipids are used to stabilize the construct, which allows a biphasic drug release profile with initial burst release continued with sustained drug release. Liposomes have shown to be successful in combating resistant pathogens. Especially, their modified designs are used to improve the potency of formulations in bacterial resistance and clearance [32]. Additionally, researchers have confirmed the feasibility of SLN and NCL as drug carriers, however, their advantages over liposomes have not been confirmed with human data [33]. Most of the research on SLN and NCL as antimicrobial carriers are still in the preclinical stage.

To date, a significant number of reports on the activity of antibiotic-conjugated polymeric nanoparticles against various infections, including those caused by drug-resistant pathogens, have been published [34]. Notably, high biocompatibility of these structures, additional to improved pharmacokinetic properties, supports the potential of these nanosystems as new tools to treat infections. Polymeric nanoparticles can be prepared from natural and synthetic polymers with the prerequisite of biocompatibility and biodegradability. In the polymeric antibacterial drug delivery systems, drug molecules can be incorporated in the internal part of the particles, on the surface of polymeric nanocarriers with covalent or non-covalent bonds, imprinted in the polymeric nanoparticles or encapsulated in the stimuli-responsive shell of polymeric nanoparticles [34]. The encapsulation route of the drug into the polymeric nanoparticle drug delivery system plays a key role in the nanocarriers' pharmacokinetic profile. The action mechanism of the polymeric nanoparticles is defined by the physicochemical properties and the composition of the particles. Polymeric nanoparticles may interact with the bacterial cell wall via passive or active targeting. Passive targeting is based on particle size and the ability of particles to disturb the structure of bacterial membrane leading to pore formation in the membrane. For active targeting of polymeric nanoparticles, the surface of polymeric nanoparticles is usually functionalized with specific antibodies and aptamer bacteriophage proteins providing specific identification for the detection of pathogens and interaction between the particles and pathogens. The reported studies reveal that both the active and passive targeting strategies to deliver antimicrobial agents with polymeric nanoparticles improve their activities compared to their free forms [35–37].

Dendrimers are highly branched macromolecules employed as antibacterial drug delivery systems. The unique properties of dendrimers, such as well-defined 3D structures, available functional groups, and their ability to mimic cell membranes, make them potential drug carriers. Both hydrophobic and hydrophilic drug molecules can be incorporated separately or at the same time into dendrimer structures. Lipophilic molecules can be incorporated inside the cavity of dendrimers, and hydrophilic agents can be covalently or physically attached to the surfaces of dendrimers. The antibacterial can be used in the building of dendrimer blocks, whereby the synthesized dendrimers themselves become potent antimicrobials. Dendrimers aid to improve the solubility, penetration, and controlled release of the drug molecules. Currently, the existing research in the design of dendrimers as antibacterial drug delivery systems also focuses on species-selective dendrimer biocide formulations. For instance, peptide, glycol, and glycopeptide dendrimer designs provide effective therapy for the bacterial infections.

An inorganic nanomaterial, in contrast to the organic materials listed above, which has also shown promise for antibacterial therapies is mesoporous silica nanoparticles (MSNs). In the design of MSN-based drug delivery systems, their advantageous characteristics (biocompatibility, high surface area, tunable particle diameter, mesoporous structure, and ease of functionalization) have been exploited. The designs with targeted and sustained release mechanisms make them powerful candidates also for antibacterial therapies. In the use of MSNs as drug delivery vehicles for antibacterial therapeutics, their surface functionality along with the size and shape are crucial parameters to improve and optimize the efficacy [38]. Their surface functionalities can be modified to target both planktonic bacteria and biofilms [39]. In recent studies, the utility of MSN for efficient antibiotic delivery [36, 40, 41] and hybrid antibacterial materials preparation by incorporating antibacterial enzymes [42], peptides [43], metal ions/particles [44], and polymers (surface modifiers) [45] to MSNs has been reported.

For rational and efficient utilization of these nanomaterial-based drug delivery systems, systematic investigation of pharmacokinetics and biodistribution should be carried out. The pharmacokinetics and biodistribution of nanoparticles are defined by their physicochemical properties [38]. Apart from their physicochemical properties, the administration routes and their elimination from the body need to be systematically evaluated. Hence, thorough evaluation of the current nanoparticle-based drug delivery systems in antibacterial therapies is important for their translation into the clinic. To date, four liposomal/lipid complex drug delivery systems for antibiotic delivery have been approved for use in human patients, including Abelcet, AmBisome, Amphotec, and Fungisome [46]. This should come as no surprise with regard that a liposomal formulation was the first nanodrug to hit the market in 1995 (Doxil®), and they have been studied since the early 1980s.

2.2. Nanomaterials as active antibacterial agents

Various types of inorganic and organic nanoparticles have been utilized as antibacterial agents. The inherent antibacterial properties of some metals and metal oxides have been known for centuries. An important advantage of antibacterial metal and metal oxide nanoparticles is that they have multiple modes of action, which is why microbes can scarcely develop resistance to them.

Among the inorganic antibacterial particles, silver nanoparticles are the most intensively investigated ones and capable to kill both Gram-positive and Gram-negative bacteria, having even shown to be effective against drug-resistant species [46]. Besides silver nanoparticles, other metal nanomaterials have also been studied for antimicrobial treatment, including gold [47], copper [48, 49], tellurium [50, 51], and bismuth [52]. Moreover, many studies have revealed the antibacterial activity of metal oxide nanomaterials, such as zinc oxide (ZnO) [53], copper oxide (CuO) [54, 55], magnesium oxide (MgO), nitric oxide (NO) [56], titanium dioxide (TiO_2) [57], aluminum oxide (Al_2O_3) [58], magnetic iron oxide (α-Fe_2O_3) [59], and cerium oxide (CeO_2) [60] nanoparticles. The toxic mode of metal and metal oxide nanoparticles against bacterial cells has been associated with ROS generation and membrane disruption [61]. According to literature findings, the release of ions is designated as the driving force behind the antimicrobial properties of antibacterial nanoparticles.

Cationic polymeric nanoparticles have been considered as promising organic antibacterial nanoparticles. They can be composed of natural or synthetic cationic polymers. The antibacterial polymeric nanoparticles kill microorganisms upon their contact with bacterial cells due to the strong interaction of their cationic surfaces with the bacterial cells [62]. The mechanisms of action have been proposed for how these cationic groups are able to disrupt the bacterial cell membrane, with some requiring hydrophobic chains of certain lengths to penetrate and burst the bacterial membrane. Moreover, different polymeric nanosized antibacterials with long-term antibacterial activity, chemically stable, and ability to bind to surfaces of interest have been reported. These include lipid nanoparticles, quaternary ammonium polyethyleneimine-incorporated polymeric nanoparticles [63, 64], chitosan [65], and self-assembled peptide nanoparticles [66]. In addition to the above-mentioned metallic and polymeric nanoparticles, carbon-based nanostructures have shown antibacterial effects. For instance, the antibacterial activity of fullerene [67] and carbon nanotubes [68, 69] (single-walled or multi-walled) derivatives have been observed. However, the antibacterial mechanism of carbon-based nanostructures is still under debate and has not received particular attention, possibly due to the difficulties of their dispersion in water, especially in case of the carbon nanotubes [70].

3. Advantages and challenges of nanotechnology-based antibacterial treatments

Promising approaches for the effective delivery of therapeutic compounds can be provided by the use of nanoparticles as drug carriers. Literature findings and clinical results have surely presented several clinical advantages of antimicrobial nanoparticles and their utilization as drug carrier systems. Antimicrobial nanoparticles are of great interest as they provide a number of benefits over free antimicrobial agents. In detail, nanocarriers can conquer the solubility and stability issues and reduce side effects [62]. With the use of nanocarriers in the delivery of drugs, combination drug therapy can be achieved by incorporating two or more drugs or different therapeutic modalities into the carrier matrix. The surface modifications can be carried out by conjugating targeting ligands on the nanocarriers that are not known by the immune system and specifically targeted to special microorganisms. Administration of antimicrobial agents using nanoparticles can increase the overall pharmacokinetics by progressing therapeutic index, extending drug circulation, and providing controlled drug release. Multiple mechanisms of action can be provided by the antibacterial nanoparticles, which prevent the development of antibacterial resistance by many pathogenic bacteria. Several routes of administration, including oral, nasal, parenteral, intraocular, and so on, can be employed with the nanotechnology-based antibacterial treatments.

The significant advantages of nanomaterials as antimicrobial agents are their modularity in design, enabling a multimodal approach that makes it especially difficult for bacteria to develop resistance mechanisms against these. Namely, a nanotechnology-based antibacterial agent can be constructed out of several components that possess antimicrobial activities in themselves, such as, for instant, be composed of an antibacterial core material (e.g. metal or metal oxide) surrounded with an antibacterial polymeric shell or coating, in which antibiotic drugs could be incorporated [71]. The core material could further be "prickly," which physically can destroy the bacterial cell wall by a "nano-piercing" process once the polymeric shell has been dissolved, leading to the disruption of bacterial integrity and lysis, as presented in a recent study by Wu et al. where zinc-doped copper oxide prickly nanoparticles exhibited high bacterial killing efficiency owing to the provided core particle nanostructure [72]. Furthermore, varying possibilities for combination therapy together with existing (commercial) antibiotics to reach synergistic effects are evident [14, 73, 74].

Although nanoparticle-based antibacterial treatments promise significant benefits and advances in addressing the key hurdles in treating infectious diseases, there are challenges in translating this exciting technology for clinical use. These include thoroughly evaluating the interactions of nanoparticles with cells, tissues, and organs, which accordingly recalibrates doses and identifies proper administration routes to obtain desired therapeutic effects. Hence, to provide a clinical translation of nanomaterials, standardized *in vitro* experimentations that will provide *in vivo* relevant data should be established [75]. In section 4, we describe existing *in vitro* methods for testing antimicrobial activity. In addition, the current methods commonly employed in testing the antibacterial and anti-biofilm activity of nanoparticles are discussed with the relevancy and pitfalls.

4. Methods to study antimicrobial activity *in vitro*

Traditionally, antimicrobial research has focused on planktonic bacteria, and there is a variety of test methods available for evaluation of antimicrobial activity against planktonic cells [76]. The Clinical and Laboratory Standards Institute (CLSI) and European Committee on Antimicrobial Susceptibility Testing (EUCAST) are the major contributors to harmonized antimicrobial susceptibility testing (AST) providing uniform procedures for testing and analysis of antibacterial activity. CLSI standards clearly define the specific and essential requirements for materials, methods, and methodologies that need to be followed without any modifications because deviations from the approved procedures might affect the experimental outcome [77]. All the critical elements for the testing, such as the culture medium, inoculum density, and incubation conditions, are listed. Unlike the antibacterial assays, standardized methods for anti-biofilm studies are scarce. No standard methods have been approved by CLSI or EUCAST for evaluation of antimicrobial activity against biofilms. Altogether, five standards (ASTM E2196, ASTM E2647, ASTM E2562, ASTM E2799, and ASTM E2871) set by the American Society for Testing and Materials (ASTM) exist, and they all are applicable as such only for *Pseudomonas aeruginosa* biofilms [78]. Moreover, only one standard is intended for susceptibility testing. Further, due to distinct phenotype and heterogeneity of biofilm bacteria, conventional *in vitro* methods used for assessment of bacterial susceptibility to antimicrobials are not appropriate for biofilm-growing bacteria [79]. Lack of standardization makes a comparison of the test results difficult, and further, can lead to the generation of conflicting data between studies since the experimental outcome is strongly dependent on the assay conditions and materials employed in the testing.

4.1. Antimicrobial susceptibility testing

In vitro susceptibility assays are performed to assess the antimicrobial susceptibility of microorganisms in order to provide efficient treatment for infections [80]. Moreover, they are used for resistance surveillance and in research laboratories to study antimicrobial activity of antimicrobial agents. Determination of the minimum inhibitory concentration (MIC) is the most widely used measure of the antimicrobial susceptibility of microorganisms. The MIC is defined as the lowest concentration of an antimicrobial (expressed as mg/L or μg/μL) required to inhibit the visible growth after overnight incubation. In addition, the minimum bactericidal concentration (MBC) is frequently used. The MBC refers to the lowest concentration of an antimicrobial that kills 99.9% of the original inoculum after a certain incubation period (**Figure 1**). These measures can be easily achieved and compared with each other when standardized methodologies and protocols are followed. A number of guidelines and standardized protocols for MIC and MBC determinations exist that include information on each step of the testing, ranging from storage and preparation of antibiotic stocks to interpretation of the results against particular microorganisms [81].

4.1.1. Dilution methods

Dilution methods are used to determine the MIC values of the antimicrobial agents. Moreover, they serve as reference methods for antimicrobial susceptibility testing. The minimum inhibitory

http://dx.doi.org/10.5772/68138

concentration can be determined both on agar (agar dilution) and in broth (broth dilution). Standards for agar and broth dilution techniques used to assess the *in vitro* susceptibility of aerobically grown bacteria are described in CLSI document M07-A9. The document covers the performance, experimental conditions, reporting of the minimal inhibitory concentration (MIC), quality control procedures, and limitations of the recommended methods.

When conducted on agar, a two-fold diluted series of the antimicrobial agent is incorporated into agar medium followed by inoculation of standardized suspension of the given organism onto the agar plate. Broth dilution can be performed in tubes (macrodilution, volume 2 ml) or in microtiter well plates (microdilution, volume ≤500 μl) containing a two-fold diluted series of antimicrobial agent prepared in the liquid growth medium that is inoculated with a standard inoculum of bacteria followed by a defined incubation period under particular conditions. After the incubation, the outcome is read based on turbidity or growth zones, and the MIC is defined. The MIC value can also be utilized to distinguish between bactericidal and bacteriostatic activities. Alternatively, when using microdilution, the MIC can be determined spectrophotometrically according to the EUCAST protocol [81]. In addition to CLSI and EUCAST standards, ISO-20776-1 standard proposes acceptable performance criteria for microdilution method. After broth dilutions, the MBC can be determined by sub-culturing the samples from tubes or wells and plating on agar to determine the number of cells (CFU/ml) after incubation for 24 h. Then, MBC is defined as the lowest concentration at which 99.9% of the final inoculum is killed (**Figure 1**). The main advantage of the dilution method is a gen-

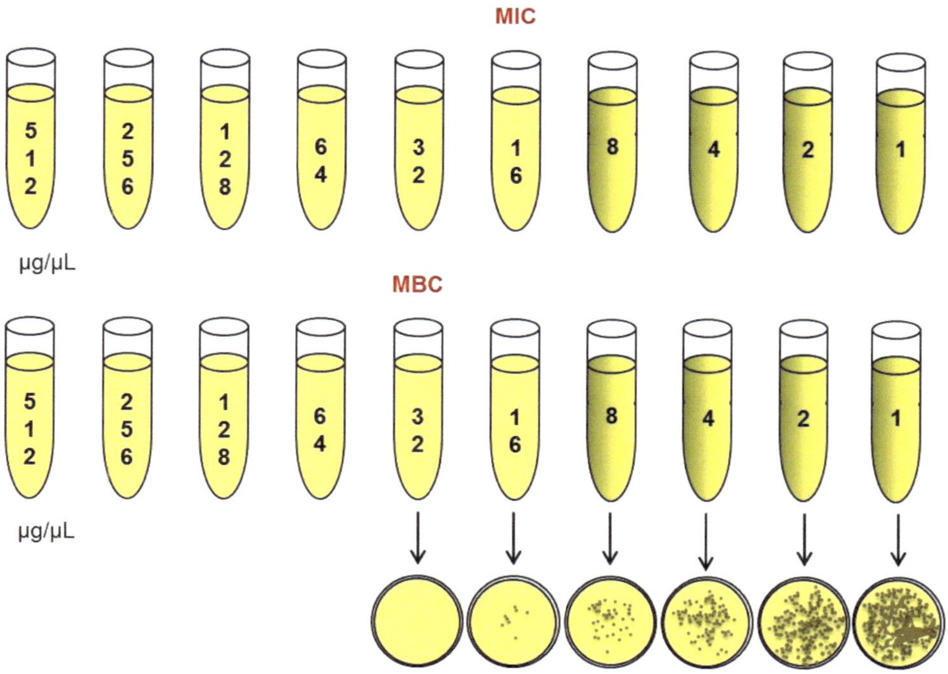

Figure 1. Minimum inhibitory concentration (MIC) versus minimum bactericidal concentration (MBC).

eration of quantitative data, the MIC value. Moreover, the assay is overall reproducible, and small amounts of antimicrobials are needed when the microdilution method is utilized. By contrast, large amounts of antimicrobial agents are needed in macrodilution testing. The main disadvantage of the method is several steps in sample preparation, which in turn, increases the possibility of errors.

4.1.2. Diffusion methods

Standards for antimicrobial disk susceptibility test are proposed in the document M02-A12 by the CLSI. Specifications for the agar (type, depth), concentration range of the test antimicrobial, concentration of the microorganism and incubation conditions (time, temperature and atmosphere) are included. Also, interpretation of the results, quality control procedures, and limitations of the methods when used for susceptibility testing of aerobically growing bacteria are described. Agar disk diffusion method is routinely applied to the *in vitro* susceptibility testing of bacteria. The Kirby-Bauer test is the most thoroughly described disk diffusion method for which interpretive standards exist (**Figure 2**). A filter paper disc impregnated with antimicrobials at different concentrations is placed on an agar plate, and the antimicrobial diffuses from the disc into the agar around the disc. Thereafter, the plate is inoculated with a standardized suspension of a microorganism followed by incubation. After the incubation period, the growth inhibition zones around the discs are measured. The diameter of the zone is depending on the antimicrobial susceptibility of microorganism. The disc diffusion test is simple to perform, but it only provides qualitative data and categorizes microorganisms as susceptible, intermediate, and resistant based on the susceptibility. Thus, it cannot be used to distinguish between bactericidal and bacteriostatic effects. Commercially available zone reader systems can be utilized to calculate an approximate MIC value by comparing the zone size and standard curve of the bacteria and antibiotic stored in an algorithm [82].

4.1.3. Combined dilution and diffusion method

The antimicrobial gradient diffusion method is based on the establishment of an antimicrobial concentration gradient in the agar medium to measure the antimicrobial susceptibility. Thin plastic test strips marked with concentration scale and impregnated with antibiotic concentration gradient are placed on agar plates that have been inoculated with a standardized

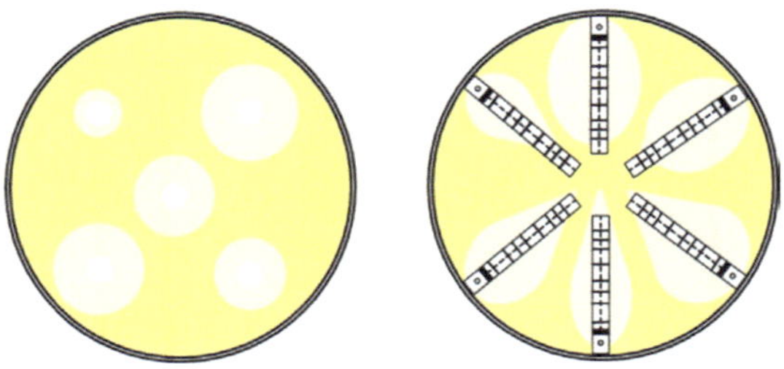

Figure 2. Disk diffusion test and antimicrobial gradient diffusion method. On the left, agar plate showing zone of inhibition by different antimicrobials of diameter of zones of inhibition refers to the susceptibility of a microorganism.

inoculum (**Figure 2**). After incubation overnight, the experimental outcome is read, and the MIC can be determined by the intersection of the lowest part of the ellipse-shaped growth inhibition area with the test strip. E-Test is the commercially available test for this purpose.

4.1.4. Time-kill test

Time-kill assay is complementary to MIC and MBC determinations. It provides information on the dynamic interaction between the antimicrobial and microorganism, thus revealing whether the antimicrobial effect is time or concentration dependent. Such activities can be investigated utilizing the standard protocol M26-A by CLSI and ASTM2315. These protocols are frequently modified. Time-kill assay is usually conducted at a concentration twice or four times the MIC. Standardized inoculum is added to a nutrient broth containing the antimicrobial at various concentrations. A sample is taken from each concentration at the inoculation time and after selected time points. Samples are serially diluted and viable plate counts are performed. The kill curves are constructed by plotting the log CFU against time. A 3-log reduction in cell counts corresponding the killing of 99.9% is considered as significant antimicrobial activity [83]. Alternatively, measurement of luminescence can be utilized to determine the time-kill relationship. Luminescence is detected by the ATP assay, in which adenosine triphosphate (ATP), an indicator for bacterial viability is quantified, and the number of viable cells is determined based on the amount of ATP. The assay uses luciferase reaction, in which luciferin is converted to oxyluciferin in the presence of molecular oxygen and ATP, and generates light by luminescence. Luminescent signal is proportional to the number of viable cells [84].

4.1.5. Methods to study antimicrobial susceptibility of biofilms

Since conventional susceptibility testing methods are not applicable for biofilms, and the MIC values do not provide a valid estimation of the antibiotic concentration needed to treat biofilm-related infections, the minimum biofilm inhibitory concentration (MBIC) has been determined instead. The MBIC determines the susceptibility of bacteria when the biofilm is forming and refers to the lowest concentration of an antimicrobial, in which no visible growth occurs after exposure to antimicrobial after the incubation period, and it can be recorded by optical reading [85]. Based upon the viable plate counts, the MBIC is defined as the lowest antimicrobial concentration in which there is no time-dependent increase in the mean number of viable cells between two exposure times [86]. Moreover, the Calgary Biofilm Device (CBD) can be used for determination of MBIC, as well as the minimum biofilm eradication concentration (MBEC), which is defined as the lowest concentration of an antimicrobial required to eradicate the established biofilms [87] along with susceptibility of planktonic bacteria (MIC) [22]. The commercially available CBD consists of 96 pegs mounted on the lid of a 96-microtiter well plate. Biofilms are first formed on the pegs for a defined time period. After the incubation period, the lid is transferred to another 96-well plate containing antimicrobials in fresh culture media at various concentrations. The MBEC is defined as the concentration of antimicrobial in which no visible growth can be detected [88]. ASTM 2799 standard describes the operational parameters required to grow, treat, sample, and analyze *Pseudomonas aeruginosa* biofilms using MBEC assay. In this assay, the experimental outcome is reported as log10 colony forming units (CFUs) per surface area, and the antimicrobial efficacy is assessed as the log10 reduction of viable cells. The experimental outcome can be evaluated using the CLSI guidelines for interpretation. However, breakpoints for resistance are not available, and those for planktonic bacteria are not applicable for biofilm bacteria [86]. Even though the ASTM protocol describes the specific experimental conditions only for *P. aeruginosa* biofilms, it can be used for other species with some modifications.

4.2. Methods to assess the anti-biofilm activity

Several assays with distinct endpoints are essential for the determination of the antimicrobial activity against biofilms. These assays rely on quantification of (i) viable cells in the biofilm, (ii) total biomass and (iii) biofilm matrix. An ideal anti-biofilm agent would target biofilm viability, biomass, and the matrix. Most of the assays are based on various staining methods. Several models have been proposed for evaluation of antimicrobial activity on biofilms. Depending on the flow of nutrients and bypass the waste products, biofilm models can be classified as closed and open systems [89]. Microtiter well plate-based assays are the most commonly used, while the Calgary biofilm device, substratum suspending reactors, and the flow cell system are the most widely used biofilm models for *in vitro* susceptibility testing [79]. Because the experimental outcome is affected by the choice of the model system, it is utmost important to select a model, in which biofilms can be formed in conditions close to *in vivo* settings.

4.2.1. Crystal violet staining

Crystal violet (Hexamethyl pararosaniline chloride, CV) assay is not only one of the oldest but also most widely used staining methods applied to biofilm quantification [90, 91]. Crystal violet is an inexpensive and basic dye that is used to measure the effects on total biomass of biofilms. Crystal violet binds indifferently to both negatively charged bacteria and polysaccharides present in the EPS. After staining, adsorbed dye is eluted in a solvent, such as acetic acid or ethanol. The amount of the dye solubilized by the solvent is measured spectrophotometrically, and it is directly proportional to biofilm biomass [92, 93]. Disadvantages of the method are shortcomings in its dynamic range, laboriousness, and low reproducibility [92]. Experimental conditions, bacterial species, concentration, and nature of the solvent used, as well as incubation time are crucial steps that affect the experimental outcome. Furthermore, the assay does not sort out living or dead cells or biofilm matrix, thus not providing any information on the number of living bacterial cells [93] and, more importantly, imprecise information on the antimicrobial activity. However, the method can be used for both Gram-negative and Gram-positive bacteria and fungi, but the optimal assay conditions, such as temperature, incubation time, and solvent, vary between species [94, 95].

4.2.2. Resazurin staining

Resazurin, also known as alamar blue (7-hydroxy-3H-phenoxazin-3-one-10-oxide), is a non-invasive, non-fluorescent dye, which is reduced to resorufin, a pink, fluorescent dye as a result of metabolically active cells and bacterial viability. Resorufin is detected spectrophotometrically to determine the viable cells [96]. Resazurin staining assay can be used to assess the antimicrobial activity based on the effects on viability of various microorganisms grown in suspensions or as biofilms [97]. However, time of resazurin reduction is species and strain specific. Consequently, the experimental conditions, such as incubation time and resazurin concentration, need to be optimized to obtain reproducible data [93, 98].

4.2.3. Fluorescein diacetate (FDA) assay

The assay is used to measure nonspecific esterase activity of viable microbial cells that converts colorless, nonfluorescent FDA to fluorescein, which is a green fluorescent compound that can be quantified fluorometrically [88]. The assay is not widely used because the dye rapidly leaks from the cells and is unstable. Moreover, hydrolysis of FDA to fluorescein in the absence of live cells and quenching of fluorescence by assay solutions may also occur under certain conditions, thus making the reliability of the assay questionable. However, the assay enables biofilm quantification without removing biofilm from the place where it has been formed, allowing the quantification of entire biofilm [84, 99].

4.2.4. Other viability assays

Biofilm viability can be assessed using tetrazolium salt reduction assay, in which tetrazolium salts, such as MTT (3-[4, 5-dimethylthiazol-2-yl]-2,5-diphenyltetrazolium bromide), XTT (2,3-bis-[2-methoxy-4-nitro-5-sulfophenyl]-2H-tetrazolium-5-carboxanilide), and MTS (3-[4, 5-dimethylthiazol-2-yl]-5-[3-carboxymethoxyphenyl]-2-[4-sulfophenyl]-2H-tetrazolium), are reduced to formazan dyes. The color change can be quantified spectrophotometrically. The amount of formazan dye produced is proportional to the number of viable cells. However, the amount of reductase enzyme produced by cells is dependent on metabolic activity [100, 101]. In other words, cells that are metabolically less active when encased in a biofilm produce also reduced amount of reductase enzyme that can lead to identification of artificially low quantities of cells when viable cells are detected using this assay. Moreover, high densities of bacteria may produce the maximum amount of formazan product leading to optical densities comparable with even higher cell densities [102].

4.2.5. Viable plate counts (colony forming units)

Viable plate counts (colony forming unit counting) are used to assess the biofilm viability based on cell counting. The assay can be used to evaluate the efficacy of antimicrobials to prevent biofilm formation or eradicate pre-formed biofilms, respectively [103]. Depending on the assay mode, bacteria and antimicrobials are added simultaneously to the microtiter well plates or biofilms are first allowed to form followed by the exposure to antimicrobial. After an incubation period, biofilms are dislodged and disaggregated. The resulting suspensions are carefully homogenized, for example, by vortexing or sonication [93]. Subsequently, suspensions are serially diluted and plated or spread on agar. Colony forming units (CFUs) per surface area or volume are counted after an incubation period. The experimental outcome can be evaluated as reduction in CFUs compared to untreated control biofilms. The method gives accurate information on bacterial viability [94]. However, it is regarded as a time-consuming and laborious methodology because of serial dilutions and plating. Additionally, special attention needs to be addressed to the detachment and disaggregation steps to avoid false negative/positive results. The complete recovery and disaggregation of biofilm need to be ensured by applying methods that do not affect viability of the biofilm cells [93, 104].

4.3. Testing of antimicrobial activities of nanoparticles

With respect to the evaluation of antimicrobial activity of nanoparticles, only one specific standard is set by the International Organization for Standardization (ISO), and it is intended for determination of the antimicrobial potency of silver nanoparticles against *Staphylococcus aureus* by measuring the release of muramic acid using gas chromatography-mass spectrometry (GC-MS) [105]. Thus, conventional methodologies for antimicrobial susceptibility testing have been not only adapted but also modified for investigation of antimicrobial activity of various nanoparticles [102]. In antimicrobial testing, experimental outcome is affected by many factors, such as solvent, inoculum preparation, type of the culture media, and incubation conditions, and these factors have been also found to be influential when testing antimicrobial activities of nanoparticles [76, 106]. For example, choice of the culture media can have a substantial impact on antimicrobial activity of nanoparticles. Media composition and its pH may affect the experimental outcome because of the impact on physicochemical properties of nanoparticles, and further, media type (solid vs. liquid) has found to profoundly influence the antimicrobial activity of nanoparticles [1, 107, 108].

Both standardized and modified microdilution and macrodilution methods have been applied to the determination of MIC and MBC values to evaluate the antimicrobial susceptibility of several microorganisms to nanoparticles [76, 109, 110–116]. Additionally, resazurin staining assay has been employed to determine the MIC. The MIC was recorded as the lowest concentration at which color change from blue to pink occurred [117]. Even though standardized antimicrobial susceptibility testing protocols can be followed, no standards describing the synthesis of nanoparticles exist. Differences in the synthesis methodology are known to impact the particle size and chemical composition of the nanoparticles, which, in turn, can further affect the antimicrobial activity and cause variability in the experimental outcome [1, 118]. Hence, the impact of such factors has to be taken into consideration when results between studies are compared.

Diffusion methods can be applied alongside dilution assays to confirm the antimicrobial susceptibility of microorganisms [108, 119, 120]. Agar disk diffusion tests performed both according to the standardized protocols and with modifications are frequently used for susceptibility assessment and evaluation of antimicrobial activity of nanoparticles [109, 121–124]. Paper disc method has been employed as an alternative to standardized single disk method [112]. Further, agar well diffusion assays have been successfully utilized for the evaluation of antimicrobial effects of nanoparticles [120, 125, 126].

Antimicrobial effects can also be determined by reading the optical density [125]. Even though measurement of optical density is a straightforward method, it is not the most suitable for measuring the activity of nanoparticles because nanoparticles as such can also interfere with the optical density [46, 127]. Viable plate counts have been frequently performed according to the various protocols to assess the antimicrobial efficacy of nanoparticles against both planktonic and biofilm-growing bacteria [109, 128, 110–112, 116]. Samples can either be spread or pipetted on agar plates followed by overnight incubation and determination of the number of CFUs [116, 111, 113–115]. Especially, when quantifying the biofilm bacteria, efficient disaggregation of samples is of great importance to avoid false positive results.

Crystal violet staining is the most widely applied staining assay to investigate the antimicrobial activity of the various nanoparticles against biofilm-growing bacteria [129, 130, 121, 116, 131, 132]. By combining assays that quantify different features of biofilms, more relevant information on the activity of nanoparticles can be obtained. In that context, effects of nanoparticles on biofilm inhibition have been studied using viable plate counts and crystal violet staining in parallel [133, 134]. Additionally, LIVE/DEAD and crystal violet staining has been combined for the same purpose [133, 134]. Crystal violet staining has been also used together with resazurin staining assay to assess the impact of nanoparticles on total biomass, including the matrix components and biofilm viability, respectively [135]. Further, crystal violet along with phenol has been applied to quantify the effects on biomass and EPS [136]. Antimicrobial agents displaying bactericidal effects have usually an impact on both viability and biomass, while antimicrobials acting like detergents affect only the biomass [137]. Further, to distinguish between bactericidal and bacteriostatic activities, LIVE and DEAD staining of bacterial biofilms can be conducted using a combination of fluorescein diacetate (FDA) and propidium iodide (PI) dyes or by commercially available LIVE/DEAD kit containing propidium iodide (PI) and SYTO9 fluorescent dyes [136, 138]. Commonly used microtiter well plate-based assays are summarized in **Table 1**.

In order to gain information on the mechanistic action of nanoparticles, antimicrobial assays can be conducted in two modes, prior to and post biofilm formation. In the pre-exposure mode, nanoparticles and bacteria are simultaneously added, whereas in post-exposure mode, biofilms are first allowed to form, followed by the exposure to nanoparticles. Crystal violet staining has been used to evaluate the impact of nanoparticles on biofilm formation and eradication, respectively [139, 140, 127, 141], and viable plate counts have been utilized in the

Assays	Endpoint	Read-out	Planktonic bacteria	Biofilm	Ref.
Measurement of optical density	Growth inhibition, MIC	Absorbance	X	X	[119, 125]
Resazurin staining assay	Viability, MIC	Fluorescence	X	X	[117, 136]
Tetrazolium salt reduction assay (MTT, XTT, MTS)	Viability	Fluorescence	X	X	[111, 132, 142, 143]
Crystal violet staining assay	Biofilm biomass, MBIC	Absorbance	–	X	[132]
Fluorescein diacetate (FDA) assay	Viability	Fluorescence	X	X	[143, 144]

Table 1. Microtiter well plate–based assays used to investigate antimicrobial activity of various nanoparticles.

investigation of the antimicrobial efficacy of nanoparticles in the prevention of biofilm formation and eradication of pre-formed biofilms [111, 112].

5. Conclusion

There is a strong demand to develop novel antimicrobial materials, and the emergence of nano-technology is creating a variety of options in this respect. Numerous nanoparticles exhibit antibacterial activity against several bacterial species. Today, nanomaterials are a promising platform to control bacterial infections in a broad range of applications. However, the absence of standardizations in testing methods leads to inconsistency in results. The foremost requirement of the assays applied to the evaluation of antimicrobial activity is reproducibility. Antimicrobial activity should be tested against various microorganisms, preferably against representatives of both Gram-negative and Gram-positive species. Moreover, a combination of several assays is preferred to confirm the activity. Several standardized methodologies exist for testing the antibacterial activity of conventional agents against planktonic bacteria. These methods are not applicable for biofilms, and further, they do not allow the prediction of the *in vivo* activity against biofilm-growing bacteria. The majority of the bacterial infections are nowadays attributed to biofilm formation, standardized test methods are urgently needed for more accurate evaluation of antimicrobial activity against biofilms. Even though nanoparticles represent a prominent approach to combat both multi-drug resistant and biofilm-related infections, lack of standardization of synthesis and testing methodologies is a significant concern. Several assays have been reported so far to test the anti-biofilm activity of nanoparticles-containing formulations. However, since antimicrobial assays are sensitive to variation in assay conditions, only standardization of these methods enables comparative analysis between studies.

Author details

Didem Şen Karaman[1]*, Suvi Manner[1], Adyary Fallarero[2] and Jessica M. Rosenholm[1]

*Address all correspondence to: dsen@abo.fi

1 Pharmaceutical Sciences Laboratory, Faculty of Science and Engineering, Åbo Akademi University, Turku, Finland

2 Pharmaceutical Biology, Division of Pharmaceutical Biosciences, Faculty of Pharmacy, University of Helsinki, Finland

References

[1] Rizzello L, Pompa PP. Nanosilver-based antibacterial drugs and devices: Mechanisms, methodological drawbacks, and guidelines. Chem Soc Rev. 2014;43(5):1501–18.

[2] Prestinaci F, Pezzotti P, Pantosti A. Antimicrobial resistance: a global multifaceted phenomenon. Pathog Glob Health. 2015;109(7):309–18.

[3] Renwick MJ, Brogan DM, Mossialos E. A systematic review and critical assessment of incentive strategies for discovery and development of novel antibiotics. J Antibiot (Tokyo). 2016;69(2):73–88.

[4] Tor Y, Fair R. Antibiotics and bacterial resistance in the 21st century. Perspect Med Chem. 2014;6:25–64.

[5] Davies D. Understanding biofilm resistance to antibacterial agents. Nat Rev Drug Discov. 2003;2(2):114–22.

[6] Worthington RJ, Richards JJ, Melander C. Small molecule control of bacterial biofilms. Org Biomol Chem. 2012;10(37):7457.

[7] Bjarnsholt T. The role of bacterial biofilms in chronic infections. APMIS. 2013;121:1–58.

[8] Michael CA, Dominey-Howes D, Labbate M. The antimicrobial resistance crisis: causes, consequences, and management. Front Public Health. 2014;2(145) 1–8.

[9] Ventola CL. The antibiotic resistance crisis: part 1: causes and threats. Pharm Ther. 2015;40(4):277–83.

[10] Luyt C-E, Bréchot N, Trouillet J-L, Chastre J. Antibiotic stewardship in the intensive care unit. Crit Care. 2014;18(480)1–12.

[11] Katz ML, Mueller LV, Polyakov M, Weinstock SF. Where have all the antibiotic patents gone? Nat Biotechnol. 2006;24(12):1529–31.

[12] Penesyan A, Gillings M, Paulsen I. Antibiotic discovery: combatting bacterial resistance in cells and in biofilm communities. Molecules. 2015;20(4):5286–98.

[13] Levy SB. Factors impacting on the problem of antibiotic resistance. J Antimicrob Chemother. 2002;49(1):25–30.

[14] Rogers SA, Huigens RW, Cavanagh J, Melander C. Synergistic effects between conventional antibiotics and 2-aminoimidazole-derived antibiofilm agents. Antimicrob Agents Chemother. 2010;54(5):2112–8.

[15] Cheng G, Dai M, Ahmed S, Hao H, Wang X, Yuan Z. antimicrobial drugs in fighting against antimicrobial resistance. Front Microbiol. 2016;7(470):1–11.

[16] Donlan RM, Costerton JW. Biofilms: survival mechanisms of clinically relevant microorganisms. Clin Microbiol Rev. 2002;15(2):167–93.

[17] Stewart PS, William Costerton J. Antibiotic resistance of bacteria in biofilms. Lancet. 2001;358(9276):135–8.

[18] Bordi C, de Bentzmann S. Hacking into bacterial biofilms: a new therapeutic challenge. Ann Intensive Care. 2011;1(1):19.

[19] Stewart PS, Franklin MJ. Physiological heterogeneity in biofilms. Nat Rev Microbiol. 2008;6(3):199–210.

[20] Landini P, Antoniani D, Burgess JG, Nijland R. Molecular mechanisms of compounds affecting bacterial biofilm formation and dispersal. Appl Microbiol Biotechnol. 2010; 86(3):813–23.

[21] Stewart PS. Mechanisms of antibiotic resistance in bacterial biofilms. Int J Med Microbiol. 2002;292(2):107–13.

[22] Ceri H, Olson ME, Stremick C, Read RR, Morck D, Buret A. The Calgary Biofilm Device: new technology for rapid determination of antibiotic susceptibilities of bacterial biofilms. J Clin Microbiol. 1999;37(6):1771–6.

[23] Piddock LJ. The crisis of no new antibiotics—what is the way forward? Lancet Infect Dis. 2012;12(3):249–53.

[24] Silver LL. Challenges of antibacterial discovery. Clin Microbiol Rev. 2011;24(1):71–109.

[25] Gao W, Thamphiwatana S, Angsantikul P, Zhang L. Nanoparticle approaches against bacterial infections: Nanoparticle against bacterial infections. Wiley Interdiscip Rev Nanomed Nanobiotechnol. 2014;6(6):532–47.

[26] Rosenholm JM, Mamaeva V, Sahlgren C, Lindén M. Nanoparticles in targeted cancer therapy: mesoporous silica nanoparticles entering preclinical development stage. Nanomed. 2012;7(1):111–20.

[27] Davis ME, Chen Z (Georgia), Shin DM. Nanoparticle therapeutics: an emerging treatment modality for cancer. Nat Rev Drug Discov. 2008;7(9):771–82.

[28] Zhang L, Gu F, Chan J, Wang A, Langer R, Farokhzad O. Nanoparticles in medicine: therapeutic applications and developments. Clin Pharmacol Ther. 2008;83(5):761–9.

[29] Nicolosi D, Scalia M, Nicolosi VM, Pignatello R. Encapsulation in fusogenic liposomes broadens the spectrum of action of vancomycin against Gram-negative bacteria. Int J Antimicrob Agents. 2010;35(6):553–8.

[30] Watarai S, Iwase T, Tajima T, Yuba E, Kono K, Sekiya Y. Application of pH-sensitive fusogenic polymer-modified liposomes for development of mucosal vaccines. Vet Immunol Immunopathol. 2014;158(1-2):62–72.

[31] Hallaj-Nezhadi S, Hassan M. Nanoliposome-based antibacterial drug delivery. Drug Deliv. 2015;22(5):581–9.

[32] Yang K, Gitter B, Rüger R, Wieland GD, Chen M, Liu X, et al. Antimicrobial peptide-modified liposomes for bacteria targeted delivery of temoporfin in photodynamic antimicrobial chemotherapy. Photochem Photobiol Sci. 2011;10(10):1593.

[33] Boukherroub R. Functionalized nanomaterials for the management of microbial infection: a strategy to address microbial drug resistance. Boston, MA: Elsevier; 2017.

[34] Michalak G, Głuszek K, Piktel E, Deptuła P, Puszkarz I, Niemirowicz K, et al. Polymeric nanoparticles – a novel solution for delivery of antimicrobial agents. Med Stud. 2016;1:56–62.

http://dx.doi.org/10.5772/68138

[35] Scott C, Abdelghany, Quinn, Ingram, Gilmore, Donnelly, et al. Gentamicin-loaded nanoparticles show improved antimicrobial effects towards Pseudomonas aeruginosa infection. Int J Nanomedicine. 2012;4053.

[36] Kavruk M, Celikbicak O, Ozalp VC, Borsa BA, Hernandez FJ, Bayramoglu G, et al. Antibiotic loaded nanocapsules functionalized with aptamer gates for targeted destruction of pathogens. Chem Commun. 2015;51(40):8492–5.

[37] Barreras US, Méndez FT, Martínez REM, Valencia CS, Rodríguez PRM, Rodríguez JPL. Chitosan nanoparticles enhance the antibacterial activity of chlorhexidine in collagen membranes used for periapical guided tissue regeneration. Mater Sci Eng C. 2016;58:1182–7.

[38] Gupta A, Landis RF, Rotello VM. Nanoparticle-based antimicrobials: surface functionality is critical. F1000Research. 2016;5:364.

[39] Hetrick EM, Shin JH, Paul HS, Schoenfisch MH. Anti-biofilm efficacy of nitric oxide-releasing silica nanoparticles. Biomaterials. 2009;30(14):2782–9.

[40] Tamanna T, Bulitta JB, Yu A. Controlling antibiotic release from mesoporous silica nano drug carriers via self-assembled polyelectrolyte coating. J Mater Sci Mater Med. 2015;26(2).

[41] Lee B-Y, Li Z, Clemens DL, Dillon BJ, Hwang AA, Zink JI, et al. Redox-triggered release of moxifloxacin from mesoporous silica nanoparticles functionalized with disulfide snaptops enhances efficacy against pneumonic tularemia in mice. Small. 2016;12(27):3690–702.

[42] Li L, Wang H. Enzyme-coated mesoporous silica nanoparticles as efficient antibacterial agents in vivo. Adv Healthc Mater. 2013;2(10):1351–60.

[43] Braun K, Pochert A, Lindén M, Davoudi M, Schmidtchen A, Nordström R, et al. Membrane interactions of mesoporous silica nanoparticles as carriers of antimicrobial peptides. J Colloid Interface Sci. 2016;475:161–70.

[44] Tian Y, Qi J, Zhang W, Cai Q, Jiang X. Facile, One-pot synthesis, and antibacterial activity of mesoporous silica nanoparticles decorated with well-dispersed silver nanoparticles. ACS Appl Mater Interfaces. 2014;6(15):12038–45.

[45] Şen Karaman D, Sarwar S, Desai D, Björk E. M, Odén M, Chakrabarti P, Rosenholm J.M, Chakraborti S. Shape engineering boosts antibacterial activity of chitosan coated mesoporous silica nanoparticle doped with silver: a mechanistic investigation. Journal of Materials Chemistry B. 2016, 4, 3292–3304.

[46] Zhu X, Radovic-Moreno AF, Wu J, Langer R, Shi J. Nanomedicine in the management of microbial infection – Overview and perspectives. Nano Today. 2014;9(4):478–98.

[47] Cui Y, Zhao Y, Tian Y, Zhang W, Lü X, Jiang X. The molecular mechanism of action of bactericidal gold nanoparticles on *Escherichia coli*. Biomaterials. 2012;33(7):2327–33.

[48] Karlsson HL, Cronholm P, Hedberg Y, Tornberg M, De Battice L, Svedhem S, et al. Cell membrane damage and protein interaction induced by copper containing nanoparticles—importance of the metal release process. Toxicology. 2013;313(1):59–69.

[49] Chatterjee AK, Chakraborty R, Basu T. Mechanism of antibacterial activity of copper nanoparticles. Nanotechnology. 2014;25(13):135101.

[50] Zare B, Faramarzi MA, Sepehrizadeh Z, Shakibaie M, Rezaie S, Shahverdi AR. Biosynthesis and recovery of rod-shaped tellurium nanoparticles and their bactericidal activities. Mater Res Bull. 2012;47(11):3719–25.

[51] Lin Z-H, Lee C-H, Chang H-Y, Chang H-T. Antibacterial activities of tellurium nanomaterials. Chem—Asian J. 2012;7(5):930–934.

[52] Claudio C-R, Chellam S. Bismuth nanoparticles: antimicrobials of broad-spectrum, low cost and safety. In: Seifalian A, de Mel A, Kalaskaer DM, editors Nanomedicine. One Central Press; 2014. Chapter 17 p. 430–438.

[53] Ansari MA, Khan HM, Khan AA, Sultan A, Azam A. Characterization of clinical strains of MSSA, MRSA and MRSE isolated from skin and soft tissue infections and the antibacterial activity of ZnO nanoparticles. World J Microbiol Biotechnol. 2012;28(4):1605–13.

[54] Meghana S, Kabra P, Chakraborty S, Padmavathy N. Understanding the pathway of antibacterial activity of copper oxide nanoparticles. RSC Adv. 2015;5(16):12293–9.

[55] Hans M, Erbe A, Mathews S, Chen Y, Solioz M, Mücklich F. Role of copper oxides in contact killing of bacteria. Langmuir. 2013;29(52):16160–6.

[56] Schairer DO, Chouake JS, Nosanchuk JD, Friedman AJ. The potential of nitric oxide releasing therapies as antimicrobial agents. Virulence. 2012;3(3):271–9.

[57] Allahverdiyev AM, Abamor ES, Bagirova M, Rafailovich M. Antimicrobial effects of TiO_2 and Ag_2O nanoparticles against drug-resistant bacteria and leishmania parasites. Future Microbiol. 2011;6(8):933–40.

[58] Ansari MA, Khan HM, Khan AA, Pal R, Cameotra SS. Antibacterial potential of Al_2O_3 nanoparticles against multidrug resistance strains of Staphylococcus aureus isolated from skin exudates. J Nanoparticle Res. 2013;15(10):1970–1982.

[59] Ismail RA, Sulaiman GM, Abdulrahman SA, Marzoog TR. Antibacterial activity of magnetic iron oxide nanoparticles synthesized by laser ablation in liquid. Mater Sci Eng C. 2015;53:286–97.

[60] Pelletier DA, Suresh AK, Holton GA, McKeown CK, Wang W, Gu B, et al. Effects of engineered cerium oxide nanoparticles on bacterial growth and viability. Appl Environ Microbiol. 2010;76(24):7981–9.

[61] Lemire JA, Harrison JJ, Turner RJ. Antimicrobial activity of metals: mechanisms, molecular targets and applications. Nat Rev Microbiol. 2013;11(6):371–84.

[62] R K, Pv K. Dendrimeric biocides—a tool for effective antimicrobial therapy. J Nanomed Nanotechnol. 2016;07(02).

[63] Beyth N, Yudovin-Farber I, Bahir R, Domb AJ, Weiss EI. Antibacterial activity of dental composites containing quaternary ammonium polyethylenimine nanoparticles against Streptococcus mutans. Biomaterials. 2006;27(21):3995–4002.

[64] Beyth N, Houri-Haddad Y, Baraness-Hadar L, Yudovin-Farber I, Domb AJ, Weiss EI. Surface antimicrobial activity and biocompatibility of incorporated polyethylenimine nanoparticles. Biomaterials. 2008;29(31):4157–63.

[65] Qi L, Xu Z, Jiang X, Hu C, Zou X. Preparation and antibacterial activity of chitosan nanoparticles. Carbohydr Res. 2004;339(16):2693–700.

[66] Gazit E. Self-assembled peptide nanostructures: the design of molecular building blocks and their technological utilization. Chem Soc Rev. 2007;36(8):1263.

[67] Lyon DY, Adams LK, Falkner JC, Alvarez PJJ. Antibacterial activity of fullerene water suspensions: effects of preparation method and particle size. Environ Sci Technol. 2006;40(14):4360–6.

[68] Kang S, Pinault M, Pfefferle LD, Elimelech M. Single-walled carbon nanotubes exhibit strong antimicrobial activity. Langmuir. 2007;23(17):8670–3.

[69] Choi J, Seo Y, Hwang J, Kim J, Jeong Y, Hwang M. Antibacterial activity and cytotoxicity of multi-walled carbon nanotubes decorated with silver nanoparticles. Int J Nanomedicine. 2014;4621.

[70] Li Q, Mahendra S, Lyon DY, Brunet L, Liga MV, Li D, et al. Antimicrobial nanomaterials for water disinfection and microbial control: potential applications and implications. Water Res. 2008;42(18):4591–602.

[71] Lam SJ, O'Brien-Simpson NM, Pantarat N, Sulistio A, Wong EHH, Chen Y-Y, et al. Combating multidrug-resistant Gram-negative bacteria with structurally nanoengineered antimicrobial peptide polymers. Nat Microbiol. 2016;1(11):16162.

[72] Wu R, Zhang H, Pan J, Zhu H, Ma Y, Cui W, et al. Spatio-design of multidimensional prickly Zn-doped CuO nanoparticle for efficient bacterial killing. Adv Mater Interfaces. 2016;3(18):1600472.

[73] Smekalova M, Aragon V, Panacek A, Prucek R, Zboril R, Kvitek L. Enhanced antibacterial effect of antibiotics in combination with silver nanoparticles against animal pathogens. Vet J. 2016;209:174–9.

[74] Li P, Li J, Wu C, Wu Q, Li J. Synergistic antibacterial effects of β-lactam antibiotic combined with silver nanoparticles. Nanotechnology. 2005;16(9):1912–7.

[75] Huh AJ, Kwon YJ. "Nanoantibiotics": a new paradigm for treating infectious diseases using nanomaterials in the antibiotics resistant era. J Controlled Release. 2011;156(2):128–45.

[76] Balouiri M, Sadiki M, Ibnsouda SK. Methods for in vitro evaluating antimicrobial activity: a review. J Pharm Anal. 2016;6(2):71–9.

[77] Jacqueline C, Caillon J. Impact of bacterial biofilm on the treatment of prosthetic joint infections. J Antimicrob Chemother. 2014;69(suppl 1):i37–40.

[78] Malone M, Goeres DM, Gosbell I, Vickery K, Jensen S, Stoodley P. Approaches to biofilm-associated infections: the need for standardized and relevant biofilm methods for clinical applications. Expert Rev Anti Infect Ther. 2017;15(2):147–56.

[79] Lebeaux D, Chauhan A, Rendueles O, Beloin C. From in vitro to in vivo models of bacterial biofilm-related infections. Pathogens. 2013;2(2):288–356.

[80] Govan JRW. Multidrug-resistant pulmonary infection in cystic fibrosis—what does "resistant" mean? J Med Microbiol. 2006;55(12):1615–7.

[81] Andrews JM. Determination of minimum inhibitory concentrations. J Antimicrob Chemother. 2001;48(suppl 1):5–16.

[82] Mermel LA, Allon M, Bouza E, Craven DE, Flynn P, O'Grady NP, et al. Clinical practice guidelines for the diagnosis and management of intravascular catheter-related infection: 2009 update by the infectious diseases society of America. Clin Infect Dis. 2009;49(1):1–45.

[83] Pankey GA, Sabath LD. Clinical relevance of bacteriostatic versus bactericidal mechanisms of action in the treatment of gram-positive bacterial infections. Clin Infect Dis. 2004;38(6):864–70.

[84] Weir E, Lawlor A, Whelan A, Regan F. The use of nanoparticles in anti-microbial materials and their characterization. The Analyst. 2008;133(7):835.

[85] Nostro A, Procopio F, Pizzimenti FC, Cannatelli MA, Bisignano G, Marino A, et al. Effects of oregano, carvacrol and thymol on *Staphylococcus aureus* and *Staphylococcus epidermidis* biofilms. J Med Microbiol. 2007;56(4):519–23.

[86] Macia MD, Rojo-Molinero E, Oliver A. Antimicrobial susceptibility testing in biofilm-growing bacteria. Clin Microbiol Infect. 2014;20(10):981–90.

[87] LaPlante KL, Mermel LA. In vitro activity of daptomycin and vancomycin lock solutions on staphylococcal biofilms in a central venous catheter model. Nephrol Dial Transplant. 2007;22(8):2239–46.

[88] Ali L, Khambaty F, Diachenko G. Investigating the suitability of the Calgary Biofilm Device for assessing the antimicrobial efficacy of new agents. Bioresour Technol. 2006; 97(15):1887–93.

[89] Coenye T, Nelis HJ. In vitro and in vivo model systems to study microbial biofilm formation. J Microbiol Methods. 2010;83(2):89–105.

[90] Stepanovic S, Vukovic D, Dakic I, Savic B, Svabic-Vlahovic M. A modified microtiter-plate test for quantification of staphylococcal biofilm formation. J Microbiol Methods. 2000;40(2):175–9.

[91] Peeters E, Nelis HJ, Coenye T. Comparison of multiple methods for quantification of microbial biofilms grown in microtiter plates. J Microbiol Methods. 2008;72(2):157–65.

[92] Junker LM, Clardy J. High-throughput screens for small-molecule inhibitors of pseudomonas aeruginosa biofilm development. Antimicrob Agents Chemother. 2007;51(10): 3582–90.

[93] Pantanella F, Valenti P, Natalizi T. Analytical techniques to study microbial biofilm on abiotic surfaces: pros and cons of the main techniques currently in use. Ann Ig Med Prev E Comunità. 2013;25(1):31–42.

[94] Pitts B, Hamilton MA, Zelver N, Stewart PS. A microtiter-plate screening method for biofilm disinfection and removal. J Microbiol Methods. 2003;54(2):269–76.

[95] O'Toole GA. Microtiter dish biofilm formation assay. J Vis Exp. 2011;47:2437.

[96] Van den Driessche F, Rigole P, Brackman G, Coenye T. Optimization of resazurin-based viability staining for quantification of microbial biofilms. J Microbiol Methods. 2014;98:31–4.

[97] Mariscal A, Lopez-Gigosos RM, Carnero-Varo M, Fernandez-Crehuet J. Fluorescent assay based on resazurin for detection of activity of disinfectants against bacterial biofilm. Appl Microbiol Biotechnol. 2009;82(4):773–83.

[98] Sandberg ME, Schellmann D, Brunhofer G, Erker T, Busygin I, Leino R, et al. Pros and cons of using resazurin staining for quantification of viable Staphylococcus aureus biofilms in a screening assay. J Microbiol Methods. 2009;78(1):104–6.

[99] Hannig C, Follo M, Hellwig E, Al-Ahmad A. Visualization of adherent micro-organisms using different techniques. J Med Microbiol. 2010;59(1):1–7.

[100] Manavathu EK, Vager DL, Vazquez JA. Development and antimicrobial susceptibility studies of in vitro monomicrobial and polymicrobial biofilm models with Aspergillus fumigatus and Pseudomonas aeruginosa. BMC Microbiol. 2014;14(1):53.

[101] Viganor L, Galdino ACM, Nunes APF, Santos KRN, Branquinha MH, Devereux M, et al. Anti-Pseudomonas aeruginosa activity of 1,10-phenanthroline-based drugs against both planktonic- and biofilm-growing cells. J Antimicrob Chemother. 2016;71(1):128–34.

[102] Webster TJ, Seil I. Antimicrobial applications of nanotechnology: methods and literature. Int J Nanomedicine. 2012;2767.

[103] Manner S, Skogman M, Goeres D, Vuorela P, Fallarero A. Systematic exploration of natural and synthetic flavonoids for the inhibition of *Staphylococcus aureus* biofilms. Int J Mol Sci. 2013;14(10):19434–51.

[104] Toté K, Berghe DV, Maes L, Cos P. A new colorimetric microtitre model for the detection of *Staphylococcus aureus* biofilms: new *S. aureus* biofilm model. Lett Appl Microbiol. 2007;46(2):249–54.

[105] ISO/TS 16550:2014 Determination of silver nanoparticles potency by release of muramic acid from Staphylococcus aureus. British Standards Institution; 2014.

[106] Kim JS, Kuk E, Yu KN, Kim J-H, Park SJ, Lee HJ, et al. Antimicrobial effects of silver nanoparticles. Nanomedicine Nanotechnol Biol Med. 2007;3(1):95–101.

[107] Sondi I, Salopek-Sondi B. Silver nanoparticles as antimicrobial agent: a case study on *E. coli* as a model for Gram-negative bacteria. J Colloid Interface Sci. 2004;275(1):177–82.

[108] Cremonini E, Zonaro E, Donini M, Lampis S, Boaretti M, Dusi S, et al. Biogenic selenium nanoparticles: characterization, antimicrobial activity and effects on human dendritic cells and fibroblasts. Microb Biotechnol. 2016;9(6):758–71.

[109] Sodagar A, Akhavan A, Hashemi E, Arab S, Pourhajibagher M, Sodagar K, et al. Evaluation of the antibacterial activity of a conventional orthodontic composite containing silver/hydroxyapatite nanoparticles. Prog Orthod. 2016;17(1).

[110] Díez-Martínez R, García-Fernández E, Manzano M, Martínez Á, Domenech M, Vallet-Regí M, et al. Auranofin-loaded nanoparticles as a new therapeutic tool to fight streptococcal infections. Sci Rep. 2016;6:19525.

[111] Takahashi C, Akachi Y, Ogawa N, Moriguchi K, Asaka T, Tanemura M, et al. Morphological study of efficacy of clarithromycin-loaded nanocarriers for treatment of biofilm infection disease. Med Mol Morphol. 2016;50(1):9–16.

[112] Barros J, Grenho L, Fontenente S, Manuel CM, Nunes OC, Melo LF, et al. *Staphylococcus aureus* and *Escherichia coli* dual-species biofilms on nanohydroxyapatite loaded with CHX or ZnO nanoparticles: DUAL-SPECIES BIOFILMS ON NANOHA-CHX OR NANOHA-ZNO. J Biomed Mater Res A. 2017;105(2):491–497.

[113] Khan ST, Ahmad J, Ahamed M, Musarrat J, Al-Khedhairy AA. Zinc oxide and titanium dioxide nanoparticles induce oxidative stress, inhibit growth, and attenuate biofilm formation activity of *Streptococcus mitis*. JBIC J Biol Inorg Chem. 2016;21(3):295–303.

[114] Krausz AE, Adler BL, Cabral V, Navati M, Doerner J, Charafeddine RA, et al. Curcumin-encapsulated nanoparticles as innovative antimicrobial and wound healing agent. Nanomed Nanotechnol Biol Med. 2015;11(1):195–206.

[115] Wu J, Xu H, Tang W, Kopelman R, Philbert MA, Xi C. Eradication of bacteria in suspension and biofilms using methylene blue-loaded dynamic nanoplatforms. Antimicrob Agents Chemother. 2009;53(7):3042–3048.

[116] Du J, Singh H, Yi T-H. Antibacterial, anti-biofilm and anticancer potentials of green synthesized silver nanoparticles using benzoin gum (Styrax benzoin) extract. Bioprocess Biosyst Eng. 2016;39(12):1923–1931.

[117] Krishnamoorthy K, Manivannan G, Kim SJ, Jeyasubramanian K, Premanathan M. Antibacterial activity of MgO nanoparticles based on lipid peroxidation by oxygen vacancy. J Nanoparticle Res. 2012;14(9).

[118] Flores CY, Miñán AG, Grillo CA, Salvarezza RC, Vericat C, Schilardi PL. Citrate-capped silver nanoparticles showing good bactericidal effect against both planktonic and sessile bacteria and a low cytotoxicity to osteoblastic cells. ACS Appl Mater Interfaces. 2013;5(8):3149–59.

[119] Ramasamy M, Lee J-H, Lee J. Potent antimicrobial and antibiofilm activities of bacteriogenically synthesized gold-silver nanoparticles against pathogenic bacteria and their physiochemical characterizations. J Biomater Appl. 2016;31(3):366–78.

[120] Liakos I, Grumezescu A, Holban A, Florin I, D'Autilia F, Carzino R, et al. Polylactic acid—lemongrass essential oil nanocapsules with antimicrobial properties. Pharmaceuticals. 2016;9(3):42.

[121] Vijayakumar S, Vaseeharan B, Malaikozhundan B, Shobiya M. Laurus nobilis leaf extract mediated green synthesis of ZnO nanoparticles: characterization and biomedical applications. Biomed Pharmacother. 2016;84:1213–22.

[122] Manju S, Malaikozhundan B, Vijayakumar S, Shanthi S, Jaishabanu A, Ekambaram P, et al. Antibacterial, antibiofilm and cytotoxic effects of Nigella sativa essential oil coated gold nanoparticles. Microb Pathog. 2016;91:129–35.

[123] Konwar A, Kalita S, Kotoky J, Chowdhury D. Chitosan–iron oxide coated graphene oxide nanocomposite hydrogel: a robust and soft antimicrobial biofilm. ACS Appl Mater Interfaces. 2016;8(32):20625–20634.

[124] Cihalova K, Chudobova D, Michalek P, Moulick A, Guran R, Kopel P, et al. *Staphylococcus aureus* and MRSA growth and biofilm formation after treatment with antibiotics and SeNPs. Int J Mol Sci. 2015;16(10):24656–24672.

[125] Ramalingam V, Rajaram R, PremKumar C, Santhanam P, Dhinesh P, Vinothkumar S, et al. Biosynthesis of silver nanoparticles from deep sea bacterium *Pseudomonas aeruginosa* JQ989348 for antimicrobial, antibiofilm, and cytotoxic activity: biosynthesis of silver nanoparticles. J Basic Microbiol. 2014;54(9):928–36.

[126] Kalishwaralal K, BarathManiKanth S, Pandian SRK, Deepak V, Gurunathan S. Silver nanoparticles impede the biofilm formation by *Pseudomonas aeruginosa* and *Staphylococcus epidermidis*. Colloids Surf B Biointerfaces. 2010;79(2):340–344.

[127] Akhil K, Jayakumar J, Gayathri G, Khan SS. Effect of various capping agents on photocatalytic, antibacterial and antibiofilm activities of ZnO nano particles. J Photochem Photobiol B. 2016;160:32–42.

[128] Sonkusre P, Singh Cameotra S. Biogenic selenium nanoparticles inhibit *Staphylococcus aureus* adherence on different surfaces. Colloids Surf B Biointerfaces. 2015;136:1051–1057.

[129] Aliasghari A, Rabbani Khorasgani M, Vaezifar S, Rahimi F, Younesi H, Khoroushi M. Evaluation of antibacterial efficiency of chitosan and chitosan nanoparticles on cariogenic streptococci: an in vitro study. Iran J Microbiol. 2016;8(2):93–100.

[130] Yeroslavsky G, Lavi R, Alishaev A, Rahimipour S. Sonochemically-produced metal-containing polydopamine nanoparticles and their antibacterial and antibiofilm activity. Langmuir. 2016;32(20):5201–12.

[131] Ahmed A, Khan AK, Anwar A, Ali SA, Shah MR. Biofilm inhibitory effect of chlorhexidine conjugated gold nanoparticles against *Klebsiella pneumoniae*. Microb Pathog. 2016;98:50–56.

[132] Shi S, Jia J, Guo X, Zhao Y, Chen D, Guo Y, et al. Reduced *Staphylococcus aureus* biofilm formation in the presence of chitosan-coated iron oxide nanoparticles. Int J Nanomed. 2016;11:6499–6506.

[133] Fazly Bazzaz BS, Khameneh B, Zarei H, Golmohammadzadeh S. Antibacterial efficacy of rifampin loaded solid lipid nanoparticles against *Staphylococcus epidermidis* biofilm. Microb Pathog. 2016;93:137–144.

[134] Giri K, Rivas Yepes L, Duncan B, Kolumam Parameswaran P, Yan B, Jiang Y, et al. Targeting bacterial biofilms via surface engineering of gold nanoparticles. RSC Adv. 2015;5(128):105551–105559.

[135] Boda SK, Broda J, Schiefer F, Weber-Heynemann J, Hoss M, Simon U, et al. Cytotoxicity of ultrasmall gold nanoparticles on planktonic and biofilm encapsulated gram-positive Staphylococci. Small. 2015;11(26):3183–3193.

[136] Skogman ME, Vuorela PM, Fallarero A. Combining biofilm matrix measurements with biomass and viability assays in susceptibility assessments of antimicrobials against *Staphylococcus aureus* biofilms. J Antibiot (Tokyo). 2012;65(9):453–459.

[137] Wirth SM, Lowry GV, Tilton RD. Natural organic matter alters biofilm tolerance to silver nanoparticles and dissolved silver. Environ Sci Technol. 2012;46(22):12687–12696.

[138] Mu H, Tang J, Liu Q, Sun C, Wang T, Duan J. Potent antibacterial nanoparticles against biofilm and intracellular bacteria. Sci Rep. 2016;6:18877.

[139] Loo C-Y, Rohanizadeh R, Young PM, Traini D, Cavaliere R, Whitchurch CB, et al. Combination of silver nanoparticles and curcumin nanoparticles for enhanced anti-biofilm activities. J Agric Food Chem. 2016;64(12):2513–2522.

[140] Niemirowicz K, Piktel E, Wilczewska A, Markiewicz K, Durnaś B, Wątek M, et al. Core-shell magnetic nanoparticles display synergistic antibacterial effects against *Pseudomonas aeruginosa* and *Staphylococcus aureus* when combined with cathelicidin LL-37 or selected ceragenins. Int J Nanomed. 2016;11:5443–55.

[141] Ionescu AC, Brambilla E, Travan A, Marsich E, Donati I, Gobbi P, et al. Silver–polysaccharide antimicrobial nanocomposite coating for methacrylic surfaces reduces Streptococcus mutans biofilm formation in vitro. J Dent. 2015;43(12):1483–1490.

[142] Gahlawat G, Shikha S, Chaddha BS, Chaudhuri SR, Mayilraj S, Choudhury AR. Microbial glycolipoprotein-capped silver nanoparticles as emerging antibacterial agents against cholera. Microb Cell Factories. 2016;15:25.

[143] Suppi S, Kasemets K, Ivask A, Künnis-Beres K, Sihtmäe M, Kurvet I, et al. A novel method for comparison of biocidal properties of nanomaterials to bacteria, yeasts and algae. J Hazard Mater. 2015;286:75–84.

[144] Kar S. Development of nano mullite based mesoporous silica biocer with incorporated bacteria for arsenic remediation. Ceram Silik. 2016;60(3):1–10.

Chapter 5

Optimizing Antimicrobial Agents in Endodontics

Patricia P. Wright and Laurence J. Walsh

Additional information is available at the end of the chapter

http://dx.doi.org/10.5772/67711

Abstract

In endodontic (root canal) treatment, a multispecies bacterial and fungal infection is present in a place that is inaccessible to the host immune system and which offers physical protection from applied topical agents. All current protocols for irrigation suffer various deficits in performance, which is why further research on alternative approaches to using antimicrobial substances is warranted. This chapter examines the technical and clinical factors which influence the performance of antimicrobial biocide-based therapies used in endodontics within dental practice, addressing issues around instability of biocides, the influence of pH, the role of physical agitation and the challenge of penetration into biofilms and into confined spaces. A range of methods to overcome the challenges in performance are described, including novel solvents and vehicles for biocides, stabilizing agents, physical agitation and the use of activation protocols including the use of intense light, ultrasound and laser-generated shockwaves to improve the effect of biocides. While specific examples are given from the dental setting of endodontics, the principles have broader application to medicine and to general industry.

Keywords: biofilms, endodontics, biocides, ultrasonics, lasers, disinfection

1. Introduction

Antimicrobial agents which act as biocides have an important place in modern health care as they overcome many of the limitations of antibiotic and antifungal medicines by attacking not one but many targets, making the development of resistance through spontaneous mutation difficult or impossible. There are a wide range of biocides in common use, and their major mechanisms of attack on bacteria and fungi vary according to the agent chosen (**Table 1**). Considering situations where infection is present and which are very challenging to treat with antibiotics, the situation of the infected root canal is the focus of this chapter.

Oxidation of biomolecules: Sodium hypochlorite (NaOCl), hydrogen peroxide (H_2O_2, ozone (O_3), photoactivated disinfection (PAD).
Cell membrane damage: Chlorhexidine (CHX), phenolic agents, calcium hydroxide ($Ca(OH)_2$), functional antimicrobial peptides, nanoparticles (NPs) (including chitosan, gold (Au) and silver (Ag)).
Protein coagulation: Laser photothermal disinfection.

Table 1. Primary mechanisms of action of broad-spectrum antimicrobial agents.

In this particular clinical situation, there is a polymicrobial infection with multiple species of bacteria as well as occasionally fungi also being present in a dense biofilm that penetrates into the tubules of the dentine of the tooth root. The organisms and their products including endotoxins cause severe inflammatory reactions in the adjacent bone and soft tissues. The location of the biofilm within the root canal system of the tooth makes it inaccessible to the host immune system, while the tubules give physical protection from agents which are applied topically. The goal of endodontic (root canal) treatment is to decontaminate the entire root canal system; however, in many cases, viable bacteria remain at the end of treatment, causing ongoing inflammation [1], with accompanying pain and other complications [2]. Retreatment using either non-surgical or surgical methods is focussed on removing persisting microbial contaminants [3].

2. The rationale for irrigation with disinfectant solutions

The use of physical instrumentation alone, such as files or ultrasonics, does not give adequate debridement of the root canal system because of its complex three-dimensional shape, but creates some space to permit better permeation of irrigation solutions [4, 5]. The mainstay of microbial therapy is the application of multiple chemical agents as partners to physical debridement. Most current protocols involve copious irrigation with 2.5–6% solutions of sodium hypochlorite (NaOCl), supplemented in some cases with further treatments using 2% chlorhexidine (CHX). The antimicrobial actions of such agents are related to their concentration, viscosity and ability to wet the walls of the root canal system [6]. Clinical studies have shown that supplementary steps such as placing a paste of calcium hydroxide ($Ca(OH)_2$) as an inter-appointment medications can further reduce the microbial bioburden [7].

The properties of an ideal irrigation solution are listed in **Table 2**. None of the current agents used alone fulfil all these requirements. This is why combinations of approaches and improved formulation are such an important goal and the focus of the current discussion.

Since all current irrigation protocols suffer various deficits in performance, further research on alternative approaches to using antimicrobial substances is warranted. The purpose of this chapter is to review the various ways that antimicrobial actions of these materials can be enhanced, using combinations of products, altered vehicles, chemical activation and physical agitation.

Effectiveness: Exerts a rapid antimicrobial action, effective on both Gram-positive and Gram-negative bacteria, effective on fungi, able to penetrate into biofilms, able to penetrate into dentine tubules, easy to dispense and use clinically, easily removed from the root canal, dissolves organic tissue, dissolves smear layer, removes debris from the canal, retains activity in the presence of organic matter.
Safety: Not allergenic, not mutagenic, not toxic to human cells, compatible with periapical tissues.
Tooth considerations: Non-staining, does not react adversely with dental materials, does not weaken tooth structure.
Cost: Low cost of manufacture, easily prepared, good long-term stability and long shelf life

Table 2. Properties of an ideal disinfecting agent for endodontics.

3. Sodium hypochlorite

As a solution ranging in concentration from 0.5 to 6%, sodium hypochlorite (NaOCl) is currently the most popular irrigant used in endodontics [8]. The solutions used in dental practice for root canal treatment are a mixture of NaOCl with sodium chloride (to make the solution isotonic), sodium hydroxide (as a pH modifier) and a surfactant. They differ from domestic preparations used as disinfectants, e.g. for sanitizing baby bottles, by being more alkaline (with a pH from 10 to 12) and having the ability to dissolve vital and non-vital soft tissues [9, 10].

Altering the pH influences both the antibacterial effects and tissue-dissolving capacity of NaOCl preparations. In low pH solutions, hypochlorous acid (HOCl) predominates, which has disinfecting actions. This small neutral molecule enters bacterial cells where it oxidizes lipids and proteins, and also reacts with ferrous ions in a reaction similar to the Fenton reaction, to produce hydroxyl radicals [11]. At high pH, both hydroxyl and hypochlorite anions are present. The hypochlorite anion can destroy protein by the formation of chloramines, while the hydroxyl ion reacts with lipids in a saponification reaction to degrade them, thus increasing tissue-dissolving capabilities [12].

3.1. Advantages

NaOCl solutions have a broad antimicrobial spectrum and are effective on both bacteria and fungi such as *Candida albicans* [13, 14]. A particular challenge in endodontics is the presence of highly resistant organisms, particularly *Enterococcus faecalis*, that survive in extremes of acidic and alkaline pH and can withstand nutritional deprivation [15]. NaOCl can disrupt biofilms of *E. faecalis*, but this requires exposure for up to 5 min [16].

Solutions of NaOCl are able to dissolve necrotic soft tissue remnants [17], because it has strong but non-specific proteolytic activity [18, 19]. A further advantage is that when mixed with hydrogen peroxide (H_2O_2), effervescence is produced which can assist in physical debridement [20].

3.2. Limitations

Bacteria located on the surface are easily accessible and can readily be inactivated. Those lodged deep within the dentinal tubules are protected from contact and thus will persist in a viable state. In the absence of aggressive agitation, NaOCl has a limited ability to eliminate *E. faecalis* when in the biofilm state and deep within dentine tubules [21].

The potency of NaOCl solutions, expressed as the available chlorine in parts per million, declines over time, because the solutions degrade readily, and even more so when exposed to heat and light. This can occur because the stock solutions are kept too long or have been dispensed into clear or translucent plastic syringes which allow in extraneous light [22].

NaOCl has an unpleasant chlorine odour, and it interacts adversely with certain antibiotic medicaments containing tetracycline, as well as with chlorhexidine, causing inactivation through oxidation. These reactions also produce highly staining end products which can cause tooth discolouration [23].

NaOCl corrodes metallic instruments, which may lead to their premature failure (breakage or separation). The proteolytic actions of NaOCl cause severe irritant reactions when only small volumes of NaOCl are extruded into soft tissues, causing dramatic swelling [24]. These reactions are more severe in formulations which include sodium hydroxide as an alkalinizing agent, because of caustic actions on soft tissues [25].

3.3. Enhancement

Adding calcium hydroxide ($Ca(OH)_2$) improves the effectiveness of NaOCl in terms of reducing bacterial counts as well as endotoxin levels [26]. Another chemical means of enhancing sodium hypochlorite is the addition of detergents to lower the surface tension and enhance penetration, for example, the quaternary ammonium compounds cetrimide and benzalkonium chloride, which themselves are antibacterial agents [27–29]. Adding in surfactants offers the option of using lower and thus less toxic concentrations of NaOCl for disinfection. The same strategy of adding detergents can be applied to other antimicrobial agents to enhance their effectiveness.

The actions of NaOCl can also be enhanced by physical activation, for example, using ultrasonics or pulsed middle infrared lasers (such as Er:YAG or Er,Cr:YSGG lasers) to agitate the solution. Lasers are better than ultrasonic agitation in this regard [30, 31]. Laser activation of NaOCl also enhances the removal of soft tissue and debris from regions that are difficult to clean [32]. Laser agitation employs low average power settings and short pulse durations; thus, there is no ablation of the root structure [33]. The general principle of using physical agitation to improve the effects of antimicrobial agents also applies to solutions of hydrogen peroxide and $Ca(OH)_2$.

4. Chlorhexidine

Chlorhexidine (CHX) comes in various forms with the gluconate and the acetate being the most widely used in dentistry as disinfectants. It is typically used as a 2% solution of the gluconate form [10, 14, 34, 35] as a final flush [10, 36, 37]. CHX has a narrow spectrum for Gram-negative bacteria but is effective against most Gram-positive bacteria and also fungi. CHX is the agent of choice when there are Gram-positive resistant enterococci present in the root canal, which may be the case in retreatment situations [38].

4.1. Advantages

The actions of CHX in the root canal are strongly influenced by pH. At pH 5.5–6.0, chlorhexidine exists as a di-cation [39, 40]. This positive charge allows it to bind to negatively charged substances including the hydroxyapatite mineral of tooth structure, bacterial polysaccharides and particularly to Gram-positive bacteria [41] as well as onto the surface of biofilms [42]. Once adsorbed onto tooth structure, CHX can prevent subsequent microbial colonization on the surface [43].

CHX exerts antifungal actions against *C. albicans*, a property it shares with hydrogen peroxide and NaOCl [44]. It also has mild anti-collagenolytic activity, but no ability to dissolve necrotic tissues [42].

4.2. Limitations

While effective against key Gram-negative endodontic pathogens such as *Porphyromonas endodontalis*, *Porphyromonas gingivalis* and *Prevotella intermedia* [38], CHX suffers from problems of inherent resistance with pseudomonads and certain other Gram-negative bacteria. Resistance to CHX occurs in *Enterobacter* spp., *Pseudomonas* spp., *Proteus* spp. and *Providencia* spp. [45].

When used as an irrigant in the root canal in the absence of physical agitation, CHX has only a limited ability to eliminate *E. faecalis* because of poor penetration into the dentinal tubules [46]. Likewise, solutions of 2% CHX penetrate poorly into biofilms and are ineffective for dissolving biofilms [47].

When in contact with tissues, CHX is irritant and can delay healing [48, 49]. As with sodium hypochlorite, care must be taken to prevent accidental extrusion into soft tissues. Using a side-vented needle rather than a conventional needle can reduce the volume of fluid that is extruded during irrigation [50, 51].

Despite reported cases of allergy to CHX in medical settings, few have occurred in dental practice. IgE-mediated allergic responses to CHX manifest as redness, itching (urticaria) and swellings [52] and can progress to anaphylaxis [53]. The high propensity for allergy distinguishes CHX from all other disinfectants discussed in this chapter.

CHX suffers from moderate problems of stability as it can undergo cleavage with oxidation, both during storage and when in contact with oxidants such as hydrogen peroxide, ozone or NaOCl. The key end product of oxidative degradation of CHX is para-chloroaniline, an orange to brown coloured substance that is highly irritant and allergenic [23, 40]. As well as causing staining of tooth structure [40], para-chloroaniline may also interfere with the seal of the final root filling [54].

4.3. Enhancements

To enhance the actions of CHX, it can be combined with low concentrations of hydrogen peroxide, mixed freshly so as to limit the possible oxidation of the molecule [55]. Likewise, adding calcium hydroxide improves the antimicrobial actions of CHX [26].

Adding positively charged detergents such as cetrimide can potentiate the antibacterial actions of CHX [39, 56, 57]; however, because of its cationic nature, it cannot be mixed with anionic detergents [56], since these cause precipitation to occur.

5. Calcium hydroxide

Water-based pastes of calcium hydroxide ($Ca(OH)_2$) are the most commonly used inter-visit dressing material in endodontics [58]. When placed in the root canal, $Ca(OH)_2$ elevates the local pH, making this unfavourable for the growth of most bacteria [43, 59]. An in vivo study revealed that after 14 days of exposure to a $Ca(OH)_2$ paste, for 41 out of 44 endodontic pathogens, both the rate of detection in samples and mean bacterial counts declined [58]. $Ca(OH)_2$ both inactivates bacteria and reduces endotoxin levels [43], through release of hydroxyl (OH^-) ions over a prolonged period of time [43]. These destroy bacterial cell walls and cytoplasmic membranes by degrading fatty acids, thereby allowing the leakage of cellular components. They also inactivate bacterial enzymes. High pH levels within bacterial cells alter the charge of various organic molecules and so interfere with the transport of bacterial nutrients [60]. OH- ions also denature proteins and damage DNA [61]. Calcium ions may contribute to antibacterial actions by neutralizing negatively charged molecules [62].

5.1. Advantages

$Ca(OH)_2$ promotes the formation of dental hard tissues, a characteristic exploited in techniques such as apexification where continued formation of the root is intended. Enhanced mineralization is due to activation of alkaline phosphatase, resulting in the release of phosphate groups which then react with calcium ions [60]. While $Ca(OH)_2$ is biocompatible, the extremely irritant and caustic nature of sodium and potassium hydroxides makes these both unsuitable for use in the root canal of teeth. Neither sodium nor potassium hydroxides have mineralizing actions.

5.2. Limitations

A saturated solution or paste of $Ca(OH)_2$ in water has a pH of approximately 12.5–12.8. When placed in a tooth, buffering by dentine proteins and carbonate ions lowers the effective pH that can be achieved by water-based solutions and pastes by 1–2 pH units, making the effective pH achieved within the dentine around 10 [63]. This is a problem because *E. faecalis* can withstand a pH of 10 and thus is resistant to the effects of traditional water-based $Ca(OH)_2$ pastes [59, 64, 65]. Likewise, many water-based $Ca(OH)_2$ products exert limited antifungal activity [44].

5.3. Enhancements

While $Ca(OH)_2$ can be combined with CHX, the combination is not significantly more effective against *E. faecalis* [46]. In contrast, $Ca(OH)_2$ pastes that are enriched with ibuprofen or diclofenac become more effective [66].

$Ca(OH)_2$ has low solubility in water (0.17% by weight at room temperature), so only small amounts can be dissolved. While the dissolution characteristics can be improved marginally when nanoparticles are used, there remains an upper ceiling of pH 12.5–12.7 for $Ca(OH)_2$ in a water-based solvent, which explains why pH values of commercial products are always in this narrow range [43].

Work based on replacing water with non-aqueous solvents has shown that several biocompatible fluids are much better solvents for $Ca(OH)_2$ than water, including propylene glycol, polyethylene glycol (PEG) and glycerol. In each of these fluids, there is higher measured OH^- release, as measured using special electrodes designed for non-aqueous solvents as well as by titration [67, 68]. A preferred solvent is a mixture of two forms of polyethylene glycol (PEG), one being the foundation (PEG 400) and the other being a thickener (PEG 3350), to generate the preferred creamy consistency required for application into the root canal. This PEG blend is a potent solvent for $Ca(OH)_2$ which has a high release of OH^- ions into any water-based environment, since PEG 400 is miscible with water [69, 70]. This then translates into greater movement of OH^- ions through the roots of human teeth than water-containing $Ca(OH)_2$ pastes [70]. This can be explained by the common ion effect, with OH^- being the common ion in water.

6. Hydrogen peroxide

At a final concentration of 3–6%, H_2O_2 is a commonly used disinfectant. It generates oxygen radicals [71], of which the hydroxyl radical is the most important since it is the strongest oxidizer [72]. H_2O_2 generates effervescence which provides physical clearance of microbial deposits [73]. The elevated oxygen concentration it creates is unfavourable for the growth of strict anaerobes.

6.1. Limitations

Elevating the pH using an alkali metal hydroxide such as lithium, sodium or potassium hydroxide will accelerate the decomposition of H_2O_2 and provide an alkaline pH [72]. The limitation in this approach is that such hydroxides are inherently caustic.

6.2. Enhancements

The actions of H_2O_2 as a disinfectant can be enhanced by adding a suitable catalyst such as manganese or ferrous ions [74]. The latter can also be employed in the Fenton reaction, where the ferrous ion reacts with H_2O_2 to form hydroxyl and other radicals [74]. Blue, violet or ultraviolet light can provide photochemical activation for this reaction, enhancing further the generation of reactive oxygen species [75]. The Fenton reaction can also occur within bacterial cells, causing cell death [11]. Likewise, photolysis of H_2O_2 using 405 nm violet light combined with ultrasound activation can potentiate hydroxyl radical formation, with a synergistic antibacterial action between light and ultrasound activation [76].

The temperature of H_2O_2 and thus its rate of breakdown can be increased by using dyes that absorb the appropriate wavelength of intense light, for example, blue light into orange or yellow dyes. Titanium dioxide in forms similar to those used in sunscreens allows broad-spectrum light sources to be used to activate H_2O_2.

Mixing ozone (O_3) and H_2O_2 gives co-catalysis and enhanced effectiveness [77]. As with H_2O_2, physical agitation of solutions of O_3 in water using ultrasound improves their effectiveness [78]. When H_2O_2 is activated by ultrasonic agitation, the enhanced production of hydroxyl radicals by sonolysis causes increased bacterial killing [79]. This ultrasonic activation can be augmented with 405 nm violet light, as both activation pathways generate hydroxyl radicals in a synergistic manner [76].

The generation of oxygen gas bubbles from H_2O_2 and the associated disrupting effect of these on microbial deposits can be enhanced using middle infrared lasers which emit in the 2700–3000 nm wavelength range, such as the Er,Cr:YSGG and Er:YAG lasers. The energy from these lasers absorbs strongly into both water and H_2O_2, creating with each laser pulse bubbles of air, steam and oxygen. Shockwaves generated by bubble implosions create shear stresses on the walls of the root canal. Fluid movements at high speeds (in the order of 100 km/h) cause disruption of microbial deposits and smear layers [80, 81]. Because the streaming movements of such fluids causes most of the fluid movement being directed back towards the point of entry of the fibre into the tooth, the volume of fluids extruded by laser activation of various irrigation fluids is no more than when conventional irrigation is undertaken using syringes [50]. The frequent replacement of fluids also provides a cooling effect, so that heat does not accumulate [82, 83].

When H_2O_2 is added to pure water, the absorption curve of the mixture is left shifted from that of water alone, allowing lasers operating at wavelengths from 900 to 1100 nm to have much stronger absorption [84]. This approach allows handheld near-infrared lasers to be used for cavitation-based removal of bacterial deposits and smear layer from the root canal. There is accompanying photothermal disinfection of bacteria located deep in dentine tubules because such wavelengths have high transmission into dentine [85, 86]. When H_2O_2 is used in this clinical application, the concentration is below 6% by volume, as this is the threshold for soft tissue injury [73].

7. Phenolic agents

The use of phenolic agents such as camphorated monochlorophenol (CMCP) and other related hydrophobic antimicrobial agents such as essential oils has a long history in endodontics. Chlorination of phenols enhances their antibacterial action [61]. The antibacterial effect of CMCP is better at low pH. This agent has a high volatility, which may enhance its penetration into dentinal tubules [43].

7.1. Limitations

Many phenolic agents are hydrophobic and have low water solubility, so they are typically formulated in a hydrophobic solvent [87]. As a result, there are poor wetting of the root canal

http://dx.doi.org/10.5772/67711

walls, limited contact with biofilms in the canal and poor penetration into the water-rich environment of dentine tubules.

7.2. Enhancements

$Ca(OH)_2$ can be added to phenolic products to make them more effective antimicrobial agents, as this disrupts the biofilm matrix [88]. The contact between hydrophobic phenolic agents and the walls of the root canal can be enhanced by formulating these into a water base containing surfactants or by using water-miscible solvents such as low molecular weight forms of PEG. The proper choice of agents can provide not only disinfectant actions but also anti-inflammatory and analgesic effects. The latter are therapeutically desirable when patients present in pain from endodontic infections [69].

8. Nanoparticles

Nanoparticles (NPs) with diameters in the range of 1–100 nm have special properties, including increased chemical reactivity and a large surface area [89]. They can exist singly, in groups or in clusters. In endodontics, NPs suggested for use as adjuncts to conventional treatments include chitosan, poly(lactic-co-glycolic acid) (PLGA), silver (Ag), gold (Au) and $Ca(OH)_2$ [90]. Chitosan is derived from the chitin exoskeletons of various arthropods [91]. It is polycationic and attaches strongly to negatively charged bacterial cell walls causing their disruption, with subsequent leakage of cellular contents. Chitosan is also capable of disrupting the extracellular matrix of biofilms [92]. Low molecular weight forms of chitosan are more antimicrobial than those with high molecular weights [93]. Carboxymethyl chitosan has been included in endodontic sealers to provide antibacterial properties [94].

PLGA is a highly biocompatible copolymer of lactic and glycolic acid. Both PLGA and chitosan NPs can be loaded with photosensitizers for use in photoactivated disinfection (PAD). Functionalization of NPs enhances the action of photosensitizers such as rose bengal and methylene blue [95, 96].

NPs of metallic oxides, such as calcium oxide, magnesium oxide, zinc oxide and titanium oxide, can exert antibacterial actions by generating reactive oxygen species [92]. The high chemical reactivity and large surface area of NPs enhance these actions over their normal counterparts, as has been shown for calcium oxide [97]. NPs can also be made from $Ca(OH)_2$ [69, 98].

Of the various metals that form NPs, silver (Ag) is of particular interest because the antibacterial effects of silver are well known in endodontics. The same principles can be employed with ionic solutions of Ag compounds (such as silver fluoride) although issues of discolouration limit their clinical application. Using NPs of Ag in the 2–12 nm range can reduce problems of tooth discolouration. AgNPs may slowly release silver ions (Ag^+) which then interact with biological systems, attracting electrons away from sulphur and nitrogen atoms in the sulfhydryl and amino groups of biological molecules such as proteins, or with the nitrogenous purines and pyrimidines of DNA and RNA [99].

Various methods are known for producing AgNPs, ranging from chemical reactions to biosynthesis through to electrochemical methods [100, 101]. Solutions containing AgNPs can be generated by high-voltage ultralow-current electrolysis of water using silver electrodes, followed by irradiation of the solution to violet light (400–430 nm wavelength). This alters the properties of the AgNPs and ensures only clusters remain rather than ions [102]. The solutions of such particles are optically clear, free of Ag ions and quite stable.

8.1. Limitations

While silver, gold, tin and zinc all have potential for use in ionic or NP forms, issues of staining and long-term stability need to be addressed. There are now systems for stabilizing stannous (tin) ions using sodium hexametaphosphate, as well as potassium iodide protocols for reducing staining from ionic silver. These have been employed in preventive and restorative dentistry but are yet to be optimized for endodontics [103, 104].

Toxicity issues with NPs require further investigation. AgNPs are not cytotoxic at 25 μg/mL or lower concentrations; however, this threshold is based on cell culture studies using a mouse fibroblast cell line (L929). Thus, for safe clinical use, animal and human studies are needed to assess the toxicological profile in greater detail [105].

8.2. Enhancements

Displacement of NPs into poorly accessible regions of the root canal, such as dentine tubules, can be achieved by using ultrasonic or pulsed laser agitation of the solution.

The application of NPs in solutions for irrigation is at an early stage of development. AgNPs have been used in combination with traditional $Ca(OH)_2$ to increase the killing of *E. faecalis* in biofilms [106]. There is also the possibility of combining metal NPs with traditional antibiotics to enhance their effectiveness. The interactions of metals with antibiotics need to be optimized, so that effectiveness is enhanced rather than impaired. Tetracyclines bind metal ions by chelation, and this alters their effectiveness. Tetracyclines with their long history of use in endodontics would be a logical first place to study such interactions. It is also possible that NPs can be functionalized with photosensitizers to increase their antimicrobial effectiveness [89]. There is already evidence that AgNPs can be combined with photosensitizers in a dual-treatment approach [107].

9. Functional peptides

Custom peptides could be designed which can inactivate particular bacterial pathogens such as *E. faecalis*. A typical primary root canal infection is a mixed flora of 20–30 species of microorganisms, so the first challenge in using antimicrobial peptides is to ensure that the coverage is sufficient. The second challenge is that some common Gram-negative endodontic pathogens such as *P. endodontalis* and *P. gingivalis* produce large quantities of proteases (such as the gingipains) [108], which could readily degrade such peptides and limit their useful life as therapeutic agents. A third challenge with peptides is their higher cost of manufacture and shelf life than other options.

10. Laser-based photothermal disinfection

Penetration of near-infrared light in the wavelength range from 800 to 1100 nm through dentine is high, allowing laser light to reach bacteria embedded deep within dentine tubules [109]. Absorption of laser energy into melanin, water, porphyrins and other molecules then denatures enzymes and so kills bacteria through photothermal actions. As the heat will then be transmitted into adjacent tooth structure and periodontal ligament, the laser energy must be delivered in pulsed mode to allow cooling. Typically, the laser is activated over several passes as the fibre is withdrawn, causing irradiation of the entire root canal system [110].

Laser energy used for photothermal disinfection can also exert biostimulatory effects and so enhance healing and reduce inflammation [111, 112]. A recent study using a 980 nm diode laser showed both bacterial inactivation and biostimulatory effects [113].

11. Photoactivated disinfection

Besides the term photoactivated disinfection (PAD), the literature in this area contains many terms that refer to the same process, such as photodynamic therapy (PDT), antimicrobial PDT (aPDT), light-activated disinfection (LAD), advanced non-invasive LAD (ANILAD) and lethal laser photosensitization (LLP).

Lasers or light-emitting diodes (LEDs) used as sources of intense light cause electronic activation of photosensitizers, which then produce ROS. Examples of light and dye combinations include visible red light with methylene blue or tolonium chloride [114], visible blue light with curcumin and green light with rhodamine B dye. The release of ROS is a photodynamic process, i.e. driven by light, with no accompanying thermal effects [115]. Even though the ROS produced are extremely short lived, photosensitizers are able to exert their actions on pathogens effect because they attach directly to microbial cell walls. The porphyrins found within many facultative and strict Gram-negative anaerobes can act as endogenous photosensitizers and absorb blue light, such that external dyes are not needed. Gram-positive bacteria and fungi do not contain large amounts of porphyrins and so are much less sensitive to the effects of blue light.

Maximal activation of photosensitizers requires many parameters to be optimized other than the specific absorption of the dye matching the emission wavelength of the light source. The dye concentration must be optimized, since too little means only a low concentration of ROS will be produced. When the dye solution is too concentrated, the optical density of the solution becomes too great and the penetration of light is limited to being superficial, and the effects become photothermal rather than photodynamic. Concentrated dyes also have the potential to stain tooth structure. Adequate amounts of oxygen need to be present locally to support the reaction. For this reason, some protocols include oxygen carriers or oxidizers such as H_2O_2 to enhance the action of the photosensitizers [116]. The pH of the solvent for the dye must be optimized for the concentration and type of ROS desired, e.g. a higher pH means more perhydroxyl radicals will form with greater antimicrobial actions. Inclusion of a surfactant is important for ensuring penetration of the dye into biofilms and into difficult-to-reach areas such as dentine tubules [117]. The optical fibre delivery systems used to deliver intense light from lasers or LEDs may include special diffusers to optimize the angular distribution of light into the dye.

PAD should be used after conventional disinfection with NaOCl, but not as a complete replacement to traditional chemical disinfectants [118, 119]. It can also be used as a supplementary form of root canal disinfection in endodontic retreatment cases [120]. An animal study which used PAD as an adjunct to standard disinfection with NaOCl reported a larger reduction in periapical lesions and more periapical regeneration than standard treatment protocols [121]. While some in vivo studies have recommended PAD as a possible alternative to NaOCl [122, 123], there remains a need to standardize the protocol and optimize the light parameters [114, 124].

11.1. Enhancements

To achieve an even distribution of light across the dye solution, a range of optical fibre tips have been developed which have side-firing capabilities, giving even lateral dispersion of light because of surface patterning [125]. Pulsing the light source can enhance the production of ROS by the dye. This has been shown in studies of PAD using thick *E. faecalis* biofilms in the root canal [126]. Pulsing the light source lowers the requirements for cooling the diode laser or LED and also improves the stability of the wavelength emitted by reducing drift to longer wavelengths. Finally, ultrasonic agitation can increase the effectiveness of PAD [127]. Photosensitizer dyes can also be used in combination with NPs as discussed above, as well as with H_2O_2 [128, 129].

12. Fluorescence control of disinfection approaches

Both planktonic microorganisms and biofilms in the root canal system can be detected using fluorescence. The point where the canal is free of microbial contamination can be identified precisely, providing an endpoint to treatment [130, 131]. Fibres with special surface characteristics allowing light delivery for fluorescence excitation and light collection from fluorescence emissions have been designed, and their performance demonstrated in various situations [132–134].

12.1. Limitations

The major limitation to such systems being used is when the fluorescence properties of tooth structure have been enhanced because of incorporation of tetracycline. This causes increased fluorescence which must be corrected for [130, 131]. While few irrigants or medicaments used in modern endodontics give fluorescence emissions, certain potent oxidizers can quench (reduce) fluorescence emissions by bacteria. Using scavengers such as sodium ascorbate or sodium thiosulphate can remove such quenching actions and allow reliable fluorescence readings to be obtained.

13. Conclusions

There are a range of options available for disinfection of the root canal. Recognizing that none of the existing agents used alone aligns with all the properties of an ideal agent (as listed in **Table 2**), the effort to enhance the capabilities of the existing agents and find new approaches must continue. Major directions for the future include improved formulations such as the use of non-aqueous solvents for $Ca(OH)_2$ for medicaments to remain in the root canal between successive appointments, as well as changes to clinical treatments delivered by the dentist during endodontic treatment. The latter include agitation of water-based disinfectant irrigants with lasers, and the combination of antimicrobial NPs with optical technologies, for synergy of effects as a final high-level disinfection step before the root canal is filled. The ideal protocol should address issues of clinical time, materials cost and complexity, as well as efficacy and safety. Using fluorescence to measure levels of pathogens can provide an end-point to clinical interventions, to inform the practitioner when pathogens no longer remain. Including antimicrobial agents into dental materials used to fill or seal the root canal is a further avenue to explore, applying the principles discussed in this chapter.

Author details

Patricia P. Wright and Laurence J. Walsh*

*Address all correspondence to: l.walsh@uq.edu.au

School of Dentistry, The University of Queensland, Brisbane, Australia

References

[1] Paiva SS, Siqueira JF Jr, Rôças IN, Carmo FL, Ferreira DC, Curvelo JA, Soares RM, Rosado AS. Supplementing the antimicrobial effects of chemomechanical debridement with either passive ultrasonic irrigation or a final rinse with chlorhexidine: a clinical study. J Endod. 2012;38:1202-6.

[2] Manfredi M, Figini L, Gagliani M, Lodi G. Single versus multiple visits for endodontic treatment of permanent teeth. Cochrane Database Syst Rev. 2016;12:CD005296.

[3] Del Fabbro M, Corbella S, Sequeira-Byron P, Tsesis I, Rosen E, Lolato A, Taschieri S. Endodontic procedures for retreatment of periapical lesions. Cochrane Database Syst Rev. 2016;10:CD005511.

[4] Carrotte P. Endodontics: part 7. Preparing the root canal. Br Dent J. 2004;197:603-13.

[5] Kuzekanani M, Walsh LJ, Yousefi MA. Cleaning and shaping curved root canals: Mtwo vs ProTaper instruments, a lab comparison. Ind J Dent Res. 2009;20:268-70.

[6] Regan JD, Fleury AA. Irrigants in non-surgical endodontic treatment. J Ir Dent Assoc. 2006;52:84-92.

[7] Paiva SS, Siqueira JF Jr, Rôças IN, Carmo FL, Leite DC, Ferreira DC, Rachid CT, Rosado AS. Clinical antimicrobial efficacy of NiTi rotary instrumentation with NaOCl irrigation, final rinse with chlorhexidine and interappointment medication: a molecular study. Int Endod J. 2013;46:225-33.

[8] Shen Y, Gao Y, Lin J, Ma J, Wang Z, Haapasalo M. Methods and models to study irrigation. Endodont Topics. 2012;27:3-34.

[9] Clarkson R, Moule A, Podlich H, Kellaway R, Macfarlane R, Lewis D, Rowell J. Dissolution of porcine incisor pulps in sodium hypochlorite solutions of varying compositions and concentrations. Aust Dent J. 2006;51:245-51.

[10] Schäfer E. Irrigation of the root canal. ENDO. 2007;1:11-27.

[11] Fukuzaki S. Mechanisms of actions of sodium hypochlorite in cleaning and disinfection processes. Biocontrol Sci. 2006;11:147-57.

[12] Estrela C, Estrela CR, Barbin EL, Spanó JCE, Marchesan MA, Pécora JD. Mechanism of action of sodium hypochlorite. Braz Dent J. 2002;13:113-7.

[13] Radcliffe C, Potouridou L, Qureshi R, Habahbeh N, Qualtrough A, Worthington H, Drucker D. Antimicrobial activity of varying concentrations of sodium hypochlorite on the endodontic microorganisms *Actinomyces israelii, A. naeslundii, Candida albicans* and *Enterococcus faecalis*. Int Endod J. 2004;37:438-46.

[14] Agrawal Vineet S, Rajesh M, Sonali K, Mukesh P. A contemporary overview of endodontic irrigants: a review. J Dent App. 2014:1:105-15.

[15] Walsh LJ, Athanassiadis B. The challenge of endodontic "superbugs" in clinical practice. Australas Dent Pract. 2008;19:102-6.

[16] Clegg MS, Vertucci FJ, Walker C, Belanger M, Britto LR. The effect of exposure to irrigant solutions on apical dentine biofilms in vitro. J Endod. 2006;32:434-7.

[17] Naenni N, Thoma K, Zehnder M. Soft tissue dissolution capacity of currently used and potential endodontic irrigants. J Endod. 2004;30:785-7.

[18] Mohammadi Z. Sodium hypochlorite in endodontics: an update review. Int Dent J. 2008;58:329-41.

[19] Mohammadi Z, Shalavi S. Antimicrobial activity of sodium hypochlorite in endodontics. J Mass Dent Soc. 2013;62:28-31.

[20] Al-Ali M, Sathorn C, Parashos P. Root canal debridement efficacy of different final irrigation protocols. Int Endod J. 2012;45:898-906.

[21] Estrela C, Silva JA, de Alencar AH, Leles CR, Decurcio DA. Efficacy of sodium hypochlorite and chlorhexidine against *Enterococcus faecalis* – a systematic review. J Appl Oral Sci. 2008;16:364-8.

[22] Clarkson R, Moule A, Podlich H. The shelf-life of sodium hypochlorite irrigating solutions. Aust Dent J. 2001;46:269-76.

[23] Walsh LJ, Athanassiadis B. Endodontic aesthetic iatrodontics. Australas Dent Pract. 2007;18:62-64.

[24] Hülsmann M, Hahn W. Complications during root canal irrigation – literature review and case reports. Int Endod J. 2000;33:186-93.

[25] Zehnder M. Root canal irrigants. J Endod. 2006;32:389-98.

[26] Xavier AC, Martinho FC, Chung A, Oliveira LD, Jorge AO, Valera MC, Carvalho CA. One-visit versus two-visit root canal treatment: effectiveness in the removal of endotoxins and cultivable bacteria. J Endod. 2013;39:959-64.

[27] De Oude N. The Handbook of Environmental Chemistry. Volume 3, Part F: Anthropogenic Compounds. Heidelberg: Springer Verlag, 1992.

[28] Baron A, Lindsey K, Sidow SJ, Dickinson D, Chuang A, McPherson JC. Effect of a benzalkonium chloride surfactant–sodium hypochlorite combination on elimination of *Enterococcus faecalis*. J Endod. 2016;42:145-9.

[29] Wang Z, Shen Y, Ma J, Haapasalo M. The effect of detergents on the antibacterial activity of disinfecting solutions in dentin. J Endod. 2012;38:948-53.

[30] Ordinola-Zapata R, Bramante C, Aprecio R, Handysides R, Jaramillo D. Biofilm removal by 6% sodium hypochlorite activated by different irrigation techniques. Int Endod J. 2014;47:659-66.

[31] Peters OA, Bardsley S, Fong J, Pandher G, DiVito E. Disinfection of root canals with photon-initiated photoacoustic streaming. J Endod. 2011;37:1008-12.

[32] Guneser MB, Arslan D, Usumez A. Tissue dissolution ability of sodium hypochlorite activated by photon-initiated photoacoustic streaming technique. J Endod. 2015;41:729-32.

[33] DiVito E, Peters O, Olivi G. Effectiveness of the erbium: YAG laser and new design radial and stripped tips in removing the smear layer after root canal instrumentation. Lasers Med Sci. 2012;27:273-80.

[34] Abraham S, Raj JD, Venugopal M. Endodontic irrigants: a comprehensive review. J Pharm Sci Res. 2015;7:5-9.

[35] Chiniforush N, Pourhajibagher M, Shahabi S, Bahador A. Clinical approach of high technology techniques for control and elimination of endodontic microbiota. J Lasers Med Sci. 2015;6:139.

[36] Haapasalo M, Shen Y, Wang Z, Gao Y. Irrigation in endodontics. Brit Dent J. 2014;216:299-303.

[37] Plotino G, Cortese T, Grande NM, Leonardi DP, Di Giorgio G, Testarelli L, Gambarini G. New technologies to improve root canal disinfection. Braz Dent J. 2016;27:3-8.

[38] Vianna ME, Gomes BP, Berber VB, Zaia AA, Ferraz CC, de Souza-Filho FJ. In vitro evaluation of the antimicrobial activity of chlorhexidine and sodium hypochlorite. Oral Surg Oral Med Oral Pathol Oral Radiol Endod. 2004;97:79-84.

[39] Portenier I, Waltimo T, Ørstavik D, Haapasalo M. Killing of *Enterococcus faecalis* by MTAD and chlorhexidine digluconate with or without cetrimide in the presence or absence of dentine powder or BSA. J Endod. 2006;32:138-41.

[40] Basrani BR, Manek S, Sodhi RN, Fillery E, Manzur A. Interaction between sodium hypochlorite and chlorhexidine gluconate. J Endod. 2007;33:966-9.

[41] Rölla G, Löe H, Schiøtt CR. Retention of chlorhexidine in the human oral cavity. Arch Oral Biol. 1971;16:1109-33.

[42] Kanisavaran ZM. Chlorhexidine gluconate in endodontics: an update review. Int Dent J. 2008;58:247-57

[43] Athanassiadis B, Abbott PV, Walsh LJ. The use of calcium hydroxide, antibiotics and biocides as antimicrobial medicaments in endodontics. Aust Dent J. 2007;52(Suppl): S64-82.

[44] Ferguson JW, Hatton JF, Gillespie MJ. Effectiveness of intracanal irrigants and medications against the yeast *Candida albicans*. J Endod. 2002;28:68-71.

[45] Kampf G. Acquired resistance to chlorhexidine – is it time to establish an 'antiseptic stewardship' initiative? J Hosp Infect. 2016;94:213-27.

[46] Saatchi M, Shokraneh A, Navaei H, Maracy MR, Shojaei H. Antibacterial effect of calcium hydroxide combined with chlorhexidine on *Enterococcus faecalis*: a systematic review and meta-analysis. J Appl Oral Sci. 2014;22:356-65.

[47] del Carpio-Perochena AE, Bramante CM, Duarte MA, Cavenago BC, Villas-Boas MH, Graeff MS, Ordinola-Zapata R. Biofilm dissolution and cleaning ability of different irrigant solutions on intraorally infected dentin. J Endod. 2011;37:1134-8.

[48] Grant PB, Walsh LJ. Problems and adverse effects associated with the use of chlorhexidine in dental practice. Periodont. 1985;6:33-8.

[49] Mohammadi Z, Abbott PV. The properties and applications of chlorhexidine in endodontics. Int Endod J. 2009;42:288-302.

[50] George R, Walsh LJ. Apical extrusion of root canal irrigants when using Er:YAG and Er,Cr:YSGG lasers with optical fibers: an in vitro dye study. J Endod. 2008;34:706-8.

[51] Chou K, George R, Walsh LJ. Effectiveness of different intra-canal irrigation techniques in removing intra-canal paste medicaments. Aust Endod J. 2014;40:21-5.

[52] Nagendran V, Wicking J, Ekbote A, Onyekwe T, Garvey LH. IgE-mediated chlorhexidine allergy: a new occupational hazard? Occup Med. 2009;59:270-2.

[53] Tran C, Walsh LJ. Anaphylaxis to dental products in dental hygiene practice. Hyg Today. 2011;11:4-7.

[54] Bui TB, Baumgartner JC, Mitchell JC. Evaluation of the interaction between sodium hypochlorite and chlorhexidine gluconate and its effect on root dentin. J Endod. 2008;34:181-5.

[55] Steinberg D, Heling I, Daniel I, Ginsburg I. Antibacterial synergistic effect of chlorhexidine and hydrogen peroxide against *Streptococcus sobrinus, Streptococcus faecalis* and *Staphylococcus aureus*. J Oral Rehab. 1999;26:151-6.

[56] Russell A, Day M. Antibacterial activity of chlorhexidine. J Hospital Infect. 1993;25:229-38.

[57] Ruiz-Linares M, Aguado-Pérez B, Baca P, Arias-Moliz MT, Ferrer-Luque CM. Efficacy of antimicrobial solutions against polymicrobial root canal biofilm. Int Endod J. 2017;50:77-83.

[58] de Souza CAS, Teles RP, Souto R, Chaves MAE, Colombo APV. Endodontic therapy associated with calcium hydroxide as an intracanal dressing: microbiologic evaluation by the checkerboard DNA-DNA hybridization technique. J Endod. 2005;31:79-83.

[59] Mohammadi Z, Dummer PM. Properties and applications of calcium hydroxide in endodontics and dental traumatology. Int Endod J. 2011;44:697-730.

[60] Estrela C, Sydney GB, Bammann LL, Felippe O. Mechanism of action of calcium and hydroxyl ions of calcium hydroxide on tissue and bacteria. Braz Dent J. 1995;6:85-90.

[61] Siqueira J, Lopes H. Mechanisms of antimicrobial activity of calcium hydroxide: a critical review. Int Endod J. 1999;32:361-9.

[62] Upadya M, Shrestha A, Kishen A. Role of efflux pump inhibitors on the antibiofilm efficacy of calcium hydroxide, chitosan nanoparticles, and light-activated disinfection. J Endod. 2011;37:1422-6.

[63] Athanassiadis B, Abbott PV, George N, Walsh LJ. In vitro study of the inactivation by dentine of some antimicrobial endodontic medicaments and their bases. Aust Dent J. 2010;55:298-305.

[64] Athanassiadis B, Abbott PV, George N, Walsh LJ. An in vitro study of the antimicrobial activity of some endodontic medicaments and their bases using an agar well diffusion assay. Aust Dent J. 2009;54:141-6.

[65] Athanassiadis B, Abbott PV, George N, Walsh LJ. An in vitro study of the anti-microbial activity of some endodontic medicaments and their bases against *Enterococcus faecalis* biofilms. Aust Dent J. 2010;55:150-5.

[66] de Freitas RP, Greatti VR, Alcalde MP, Cavenago BC, Vivan RR, Duarte, MAH, Weckwerth PH. Effect of the association of nonsteroidal anti-inflammatory and antibiotic drugs on antibiofilm activity and pH of calcium hydroxide pastes. J Endod. 2017;43:131-4.

[67] Walsh LJ, Athanassiadis B. Non-aqueous solvents influence pH of calcium hydroxide products. J Dent Res. 2014;94(Spec Iss C):193435.

[68] Walsh LJ. Athanassiadis B. PEG 400 solvent alters pH in calcium hydroxide endodontic medicaments. J Dent Res. 2016;96(Spec Iss C):1705.

[69] Walsh LJ, Athanassiadis B. Alkaline compositions and their dental and medical use. US patent publication US 2014/0322144-A1.

[70] Teoh YY, Athanassiadis B, Walsh LJ. The influence of aqueous and PEG 400 solvent vehicles on hydroxyl ion release from calcium hydroxide medicaments. Internat Dent. 2016;11:42-50.

[71] Ikai H, Nakamura K, Shirato M, Kanno T, Iwasawa A, Sasaki K, Kohno M. Photolysis of hydrogen peroxide, an effective disinfection system via hydroxyl radical formation. Antimicrob Agent Chemother. 2010;54:5086-91.

[72] De Moor RJG, Verheyen J, Diachuk A, Verheyen P, Meire MA, De Coster PJ, Walsh LJ. Insight in the chemistry of laser-activated dental bleaching. Scientif World J. 2015;2015:650492.

[73] Walsh LJ. Safety issues relating to the use of hydrogen peroxide in dentistry. Aust Dent J. 2000;45:257-69.

[74] Neyens E, Baeyens J. A review of classic Fenton's peroxidation as an advanced oxidation technique. J Hazard Mater. 2003;98:33-50.

[75] Walsh LJ. Photoactivated disinfection (PAD) and endodontics. Australas Dent Pract. 2003;14:88.

[76] Ibi H, Hayashi M, Yoshino F, Tamura M, Yoshida A, Kobayash Y, Ogiso B. Bactericidal effect of hydroxyl radicals generated by the sonolysis and photolysis of hydrogen peroxide for endodontic applications. Microbial Pathogen. 2017;103:65-70.

[77] Walsh LJ. 'Radical' progress: ozone-boosted in-office power whitening. ADA News Bull. 2006,341:26-8.

[78] Case PD, George R, Bird PS, Walsh LJ. Treatment of root canal biofilms of *Enterococcus faecalis* with ozone gas and passive ultrasound activation. J Endod. 2012;38:523-6.

[79] Kobayashi Y, Hayashi M, Yoshino F, Tamura M, Yoshida A, Ibi H, Ogiso B. Bactericidal effect of hydroxyl radicals generated from a low concentration hydrogen peroxide with ultrasound in endodontic treatment. J Clin Biochem Nutr. 2014;54:161-5.

[80] George R, Meyers IA, Walsh LJ. Laser activation of endodontic irrigants with improved conical laser fiber tips for removing smear layer in the apical third of the root canal. J Endod. 2008;34:1524-7.

[81] George R, Chan K, Walsh LJ. Laser-induced agitation and cavitation from proprietary honeycomb tips for endodontic applications. Lasers Med Sci. 2015;30:1203-8.

[82] Hmud R, Kahler WA, Walsh LJ. Temperature changes accompanying infrared diode laser treatment of wet canals. J Endod. 2010;36:908-11.

[83] George R, Walsh LJ. Thermal effects from modified endodontic laser tips used in the apical third of root canals with erbium-doped yttrium aluminium garnet and erbium,

chromium-doped yttrium scandium gallium garnet lasers. Photomed Laser Surg. 2010;28:161-5.

[84] Hmud R, Kahler WA, George R, Walsh LJ. Cavitational effects in aqueous endodontic irrigants generated by near infrared lasers. J Endod. 2010;36:275-8.

[85] George R, Walsh LJ. Laser fiber-optic modifications and their role in endodontics. J Laser Dent. 2012;20:24-30.

[86] Lagemann M, Chai L, George R, Walsh LJ. Activation of ethylenediaminetetraacetic acid by a 940 nm diode laser for enhanced removal of smear layer. Aust Endod J. 2014;40:72-5.

[87] Walsh LJ. The role of natural and synthetic phenolic compounds in home oral health care. Australas Dent Pract. 2004;6:22-5.

[88] Lei L, Shao M, Yang Y, Mao M, Yang Y, Hu T. Exopolysaccharide dispelled by calcium hydroxide with volatile vehicles related to bactericidal effect for root canal medication. J Appl Oral Sci. 2016;24:487-95.

[89] Shrestha A, Kishen A. Antibacterial nanoparticles in endodontics: a review. J Endod. 2016;42:1417-26.

[90] Samiei M, Farjami A, Dizaj SM, Lotfipour F. Nanoparticles for antimicrobial purposes in endodontics: a systematic review of in vitro studies. Mater Sci Eng C Mater Biol Appl. 2016;58:1269-78.

[91] Reece JB, Meyers N, Urry LA, Cain ML, Wasserman SA, Minorsky PV, Jackson RB, Cooke BN. Campbell Biology, 8th edn. San Francisco: Pearson Benjamin Cummings, 2009.

[92] Kishen A. Nanoparticles for endodontic disinfection. In: Kishen A (Ed.) Nanotechnology in Endodontics: Current and Potential Clinical Applications. Switzerland: Springer International Publishing, 2015. pp. 97-119.

[93] de Paz LEC, Resin A, Howard KA, Sutherland DS, Wejse PL. Antimicrobial effect of chitosan nanoparticles on *Streptococcus mutans* biofilms. Appl Env Microbiol. 2011;77:3892-5.

[94] del Carpio-Perochena A, Kishen A, Shrestha A, Bramante CM. Antibacterial properties associated with chitosan nanoparticle treatment on root dentin and 2 types of endodontic sealers. J Endod. 2015;41:1353-8.

[95] Shrestha A, Kishen A. Antibiofilm efficacy of photosensitizer-functionalized bioactive nanoparticles on multispecies biofilm. J Endod. 2014;40:1604-10.

[96] Pagonis TC, Chen J, Fontana CR, Devalapally H, Ruggiero K, Song X, Yamazaki H. Nanoparticle-based endodontic antimicrobial photodynamic therapy. J Endod. 2010;36:322-8.

[97] Louwakul P, Saelo A, Khemaleelakul S. Efficacy of calcium oxide and calcium hydroxide nanoparticles on the elimination of *Enterococcus faecalis* in human root dentin. Clin Oral Invest. 2016; Apr 30. [Epub ahead of print].

[98] Samanta A, Chanda DK, Das PS, Ghosh J, Mukhopadhyay AK, Dey A. Synthesis of nano calcium hydroxide in aqueous medium. J Am Ceram Soc. 2016;99:787-95.

[99] Chernousova S, Epple M. Silver as antibacterial agent: ion, nanoparticle, and metal. Angew Chem Int Ed Eng. 2013;52:1636-53.

[100] Maddinedi SB, Mandal BK, Anna KK. Environment friendly approach for size controllable synthesis of biocompatible silver nanoparticles using diastase. Environ Toxicol Pharmacol. 2016;49:131-6.

[101] AbdelRahim K, Mahmoud SY, Ali AM, Almaary KS, Mustafa AE, Husseiny SM. Extracellular biosynthesis of silver nanoparticles using *Rhizopus stolonifer*. Saudi J Biol Sci. 2017;24:208-16.

[102] Laroo H. Colloidal nano silver- its production method, properties, standards and its bio-efficacy as an inorganic antibiotic. J Phys Chem Biophys. 2013;3:130.

[103] Knight GM, McIntyre JM, Craig GG, Mulyani. Ion uptake into demineralized dentine from glass ionomer cement following pretreatment with silver fluoride and potassium iodide. Aust Dent J. 2006;51:237-41.

[104] Deutsch A. An alternate technique of care using silver fluoride followed by stannous fluoride in the management of root caries in aged care. Spec Care Dentist. 2016;36:85-92.

[105] Takamiya AS, Monteiro DR, Bernabé DG, Gorup LF, Camargo ER, Gomes-Filho JE, Oliveira SH, Barbosa DB. In vitro and in vivo toxicity evaluation of colloidal silver nanoparticles used in endodontic treatments. J Endod. 2016;42:953-60.

[106] Zhang FH, Li M, Wei ZJ, Zhao B. The effect of a combined nanoparticulate/calcium hydroxide medication on the biofilm of *Enterococcus faecalis* in starvation phase. Shanghai Kou Qiang Yi Xue. 2016;25:11-5.

[107] Afkhami F, Akbari S, Chiniforush N. *Entrococcus faecalis* elimination in root canals using silver nanoparticles, photodynamic therapy, diode laser, or laser-activated nanoparticles: an in vitro study. J Endod. 2016 Dec 24. pii: S0099-2399(16)30582-9.

[108] Pathirana RD, O'Brien-Simpson NM, Veith PD, Riley PF, Reynolds EC. Characterization of proteinase-adhesin complexes of *Porphyromonas gingivalis*. Microbiol. 2006;152:2381-94.

[109] Gutknecht N, van Gogswaardt D, Conrads G, Apel C, Schubert C, Lampert F. Diode laser radiation and its bactericidal effect in root canal wall dentin. J Clin Laser Med Surg. 2000;18:57-60.

[110] Beer F, Buchmair A, Wernisch J, Georgopoulos A, Moritz A. Comparison of two diode lasers on bactericidity in root canals – an in vitro study. Lasers Med Sci. 2012;27:361-4.

[111] Walsh LJ. The current status of low level laser therapy in dentistry. Part 1. Soft tissue applications. Aust Dent J. 1997;42:247-54.

[112] Walsh LJ. The current status of low level laser therapy in dentistry. Part 2. Hard tissue applications. Aust Dent J. 1997;42:302-6.

[113] Caccianiga G, Baldoni M, Ghisalberti CA, Paiusco A. A preliminary in vitro study on the efficacy of high-power photodynamic therapy (HLLT): comparison between pulsed diode lasers and superpulsed diode lasers and impact of hydrogen peroxide with controlled stabilization. BioMed Res Int. 2016;2016:1386158.

[114] Diogo P, Goncalves T, Palma P, Santos JM. Photodynamic antimicrobial chemotherapy for root canal system asepsis: a narrative literature review. Int J Dent. 2015;2015:269205.

[115] Walsh LJ. The current status of laser applications in dentistry. Aust Dent J. 2003;48:146-55.

[116] George S, Kishen A. Augmenting the antibiofilm efficacy of advanced noninvasive light activated disinfection with emulsified oxidizer and oxygen carrier. J Endod. 2008;34:1119-23.

[117] Lee MT, Bird PS, Walsh LJ. Photo-activated disinfection of the root canal: a new role for lasers in endodontics. Aust Endod J. 2004;30:93-8.

[118] Siddiqui SH, Awan KH, Javed F. Bactericidal efficacy of photodynamic therapy against *Enterococcus faecalis* in infected root canals: a systematic literature review. Photodiagnosis Photodyn Ther. 2013;10:632-43.

[119] Arneiro RA, Nakano RD, Antunes LA, Ferreira GB, Fontes K, Antunes LS. Efficacy of antimicrobial photodynamic therapy for root canals infected with *Enterococcus faecalis*. J Oral Sci. 2014;56:277-85.

[120] Tennert C, Feldmann K, Haamann E, Al-Ahmad A, Follo M, Wrbas KT, Hellwig E, Altenburger MJ. Effect of photodynamic therapy (PDT) on *Enterococcus faecalis* biofilm in experimental primary and secondary endodontic infections. BMC Oral Health. 2014;14:132.

[121] Silva LAB, Novaes AB, de Oliveira RR, Nelson-Filho P, Santamaria M, Silva RAB. Antimicrobial photodynamic therapy for the treatment of teeth with apical periodontitis: a histopathological evaluation. J Endod. 2012;38:360-6.

[122] Bonsor S, Nichol R, Reid T, Pearson G. An alternative regimen for root canal disinfection. Brit Dental J. 2006;201:101-5.

[123] Bonsor S, Nichol R, Reid T, Pearson G. Microbiological evaluation of photo-activated disinfection in endodontics (an in vivo study). Brit Dent J. 2006;200:337-41.

[124] Trindade AC, De Figueiredo JAP, Steier L, Weber JBB. Photodynamic therapy in endodontics: a literature review. Photomed Laser Surg. 2015;33:175-82.

[125] Walsh LJ, George R. Surface structure modification. US patent 8,977,085. 2015.

[126] Hislop J, Bird PS, Walsh LJ. Photoactivated disinfection of *Enterococcus faecalis* in biofilms. Aust Dent J. 2010;55 (Suppl):S22.

[127] Tennert C, Drews AM, Walther V, Altenburger MJ, Karygianni L, Wrbas KT, Hellwig E, Al-Ahmad A. Ultrasonic activation and chemical modification of photosensitizers enhances the effects of photodynamic therapy against *Enterococcus faecalis* root-canal isolates. Photodiagnosis Photodyn Ther. 2015;12:244-51.

[128] George S, Kishen A. Photophysical, photochemical, and photobiological characterization of methylene blue formulations for light-activated root canal disinfection. J Biomed Optics. 2007;12:034029.

[129] Garcez AS, Núnez SC, Baptista MS, Daghastanli NA, Itri R, Hamblin MR, Ribeiro MS. Antimicrobial mechanisms behind photodynamic effect in the presence of hydrogen peroxide. Photochem Photobiol Sci. 2011;10:483-490.

[130] Sainsbury AL, Bird PS, Walsh LJ. DIAGNOdent laser fluorescence assessment of endodontic infection. J Endod. 2009;35:1404-7.

[131] Ho QV, George R, Sainsbury AL, Kahler WA, Walsh LJ. Laser fluorescence assessment of the root canal using plain and conical optical fibers. J Endod. 2010;36:119-22.

[132] Walsh LJ. Applications of laser fluorescence for diagnosis of bacterial infections in the root canal. Australas Dent Pract. 2010;21:54-6.

[133] Shakibaie F, George R, Walsh LJ. Applications of laser-induced fluorescence in dentistry. Int J Dent Clin. 2011;3:26029.

[134] Walsh LJ. Fluorescence applications in dentistry: current status and future prospects. Australas Dent Pract. 2013;24:62-4.

Chapter 6

Determining the Antibiotic Resistance of Bacterial Pathogens in Sexually Transmitted Diseases

Vică Mihaela Laura, Matei Horea Vladi and
Siserman Costel Vasile

Additional information is available at the end of the chapter

http://dx.doi.org/10.5772/67871

Abstract

Sexually transmitted diseases (STD) are among the most common infections worldwide. These bacterial infections have spread predominantly in the developing/underdeveloped countries, the most common being syphilis, gonorrhea and those induced by *Chlamydia trachomatis, Ureaplasma urealyticum* or *Mycoplasma* spp. Due to extensive usage of antibiotics in the recent past, these bacteria developed resistance to those commonly used for treatment, such resistant strains becoming a public health problem in a number of countries. It is well documented that bacterial STD agents are difficult to detect using standard culture media because these methods require special conditions and adequate nutrients. Antimicrobial susceptibility testing is, therefore, difficult to obtain in such cases. In recent years, genetic tests have been frequently employed in STD diagnosis. The study of genes that induce resistance to antibiotics using DNA isolated from these bacteria may prove to be a viable alternative. Genetic methods enable the DNA extraction from different biological samples, and both the presence of the bacteria and their resistance to one or more antibiotics can be determined from a single DNA sample. By studying the genes that induce antibiotic resistance and the plasmids that transfer such genes, the mechanism that leads to antibiotic resistance can be elucidated.

Keywords: STD pathogens, antibiotic resistance, genes

1. Introduction

Sexually transmitted diseases (STD) are caused by a wide variety of bacteria, viruses or parasites transferred from one human being to another primarily by vaginal, anal or oral sexual contact. Different STD can coexist or be simultaneously transmitted, the presence of any such infection increasing the risk of contracting other STD types. Prolonged, untreated infections with these pathogens can induce anexitis, endometritis, pelvic peritonitis or spontaneous abortions (in women) and urethritis, prostatitis or epididymitis (in men), respectively.

Reasons for STDs' recurrence in some sexually active women include the anatomy of the female sex, multiple partners or nonusage of condoms [1]. Such infections present an increasingly serious threat to global public health, especially for mothers or babies, causing severe complications such as cancer of the cervix, spontaneous abortions, premature births, low birth weight or infertility [2]. A number of studies concluded that ulcerative and nonulcerative STDs significantly increase HIV transmission, and drug resistance is complicating the fight against HIV [3]. The presence in the genital tract of such microbial pathogens may be due to a poor hygiene or risky behaviors as mentioned above [4].

Microorganisms make use of a wide range of mechanisms to resist to antibiotics and survive their attack. The emergence of antibiotic resistance is a natural biological phenomenon that occurs as a reaction to the use of antibiotics. Antibiotic resistance results from the bacteria ability to resist to antibiotics as they develop and multiply in media with high antibiotic concentration.

The main causes of antibiotic resistance are their abusive usage and the spread of resistant bacterial strains or genes bearing information able to induce resistance. There are multiple mechanisms leading to antibiotic resistance, one involving resistance plasmids carrying genes responsible for resistance to antibiotics [5, 6]. The phenomenon of antibiotic resistance can also develop via mutations or by acquiring resistance genes from other resistant bacteria [7–9].

Resistance to macrolides observed in the STD strains is often associated with mutations in the ribosomal protein genes [10, 11]. It is assumed that mutations in a specific region of the 23S rRNA play a major role in inducing resistance to macrolides as a result of their interaction with the ribosomes [12, 13].

Treatment failure because of bacterial resistance to the major groups of antibiotics as a result of extensive usage became a serious threat to human health. Increased usage of nucleic acid amplification tests (NAATs) in the diagnosis of STD leads to the elimination of culture-based techniques, depleting the number of live strains available for resistance testing. A lack of routine test-of-cure makes it impossible to distinguish treatment failure from reinfection [14]. The cost of health care for patients with resistant infections is higher due to longer duration of illness, additional tests and use of more expensive drugs.

It is obvious that the microorganisms' antimicrobial resistance is a dynamic phenomenon which points out the need for updated prevalence and susceptibility data from vast geographic areas being available for relevant institutions [4].

2. Most common bacterial STD agents and their resistance to antibiotics

STD infections are induced by a diversity of bacterial pathoges, including *Neisseria gonorrhoeae*, *Mycoplasma hominis*, *Mycoplasma genitalium* and *Ureaplasma urealyticum*. However, *Chlamydia trachomatis* is the leading cause of sexually transmitted bacterial infections worldwide and, in developing countries, the leading cause of preventable blindness [15].

C. trachomatis strains are obligate intracellular bacteria presenting a unique biphasic developmental cycle. Following the endocytosis of the small infectious bodies into the host cell and their inclusion into specialized vesicles, *C. trachomatis* strains differentiate into metabolically active reticular bodies and replicate until the inclusion fills the host cell. At this stage, the reticular bodies differentiate into elementary bodies and are being released from the cell to continue the infection [14].

Infection further leads to diseases such as cervicitis, pelvic inflammatory disease, urethritis, proctitis, lymphogranuloma venereum, trachoma or reactive arthritis. Transmission of STD pathoges is largely uncontrolled as over 70% of the women and 50% of the men infected with *C. trachomatis* are asymptomatic. "Silent" chronic infection has been recognized as a significant cause of infertility in women. Symptomatic individuals seeking medical care often do not benefit from tests for initial diagnosis or cure, largely due to inherent costs. Empirical antibiotics treatment is hastening the development of drug resistance, as has already occurred for a closely related species, *Chlamydia suis*, in which a tetracycline transposon was acquired from another pathogens residing in the guts of pigs [16].

C. trachomatis is a Gram-negative parasite causing a wide range of inflammations in the urogenital tract. Clinical isolates exhibiting resistance to azithromycin and associated with recurrent infections have been described previously [17]. While several reports of human chlamydial infections that do not respond to tetracycline or doxycycline were produced, no human pathogenic chlamydial strains that demonstrate stable tetracycline resistance have yet been isolated [17, 18].

Plasmids are small circular self-replicating DNA particles coexisting in the bacterial cells that are capable of exporting genetic information from the bacterial chromosome. They can transfer genes usually not present in the bacterial chromosome. In prokaryotic organisms, plasmids usually code for genes that facilitate survival of the bacteria in the environment. In chlamydia, apart from their participation in glycogen synthesis and encoding proteins responsible for their virulence, their role has yet to be clearly recognized. Resistance plasmids carry genes responsible for antibiotics resistance. Plasmids and bacteriophages can be employed in the diagnostics of chlamydioses. Plasmids are already used for detection of chlamydial infections. Bacteriophages could be used as therapeutic agents, potentially replacing antibiotics to address the increasing problem of antibiotic resistance among chlamydia. The easy transfer of infectious elementary bodies into other cells can prove to be a useful biotechnological tool in the treatment of chlamydioses, particularly in the case of concealed and latent infections, principally at the time of chlamydial resistance to antibiotics used by choice of treatment (tetracyclines) [5]. According to some authors, chlamydial plasmids are not conjugative plasmids, and they do not code antibiotic resistance and do not share the capacity for integration, except for the integrative plasmid revealed in *C. suis*, which is integrated in a genomic island together with some insertion sequences in the bacterial chromosome that encodes genes of resistance to tetracycline (*tet*C) [16].

Chlamydial infections have high recurrence rates among sexually active populations [19]. However, whether this recurrence rate for the disease is due to reinfection or to persistent infection with the same organism has been difficult to determine [20]. Immunity to chlamydial infections is type specific; once the initial infection was resolved, reinfection is believed to result from exposure to chlamydial strains differing in type from the initial infecting strain [21]. In contrast, persistent infections are considered those in which *Chlamydia* has entered a non-infectious state that is metabolically quiescent.

Chlamydial infections are very likely to recur even when appropriate drug therapy is in place. Most clinical failures occur due to reinfection or relapse as the bacterial phenotype deviates to persistent, non replicating antibiotic-resistant types able to revert to the typical reticulate body phenotype once treatment is complete [22, 23]. Administration of tetracyclines and macrolides, impeding bacterial translation by binding to the 30S or 50S ribosomal subunits, is viewed as the recommended first-line therapeutic regimen for chlamydial infections [24]. Clinical isolates from patients with recurrent *C. trachomatis* infection were documented to have significant resistance against macrolides [25, 7].

Other frequently diagnosed STD worldwide is gonorrhoea, exhibiting a recent rise in its global incidence. *N. gonorrhoeae* are Gram-negative facultative intracellular bacteria accounting for about three-quarters of urethral discharge cases among South Africa men presenting to primary healthcare clinics [8]. *Chlamydia* and gonorrhoea share several aspects of their disease outcomes, both being able to cause pelvic inflammatory disease and epithelial scarring which can lead to infertility. On the other hand, they can also be asymptomatic. However, *N. gonorrhoeae* infections tend to be more acute, with symptoms including purulent discharge and acute local inflammation. Rashes and septic arthritis may occur if the bacterium disseminates through the host [14].

While not required for individual patient management, laboratory investigations are instrumental in the choice of antimicrobial agents. Due to a high prevalence of fluoroquinolone-resistant *N. gonorrhoeae* strains observed in the last decade, first-line therapy was changed to single-dose oral cefixime. However, ciprofloxacin—the first line treatment option until about 2008—is still recommended for the treatment of presumptive gonorrhoea in patients with a history of severe allergy to cephalosporins. The increasing prevalence of *N. gonorrhoeae* antimicrobial resistance has become a global public health problem as lesser abilities to develop *N. gonorrhoeae* cultures and to perform antimicrobial susceptibility testing make monitoring antimicrobial resistance rather difficult. A significant advantage to public health programs and selective patient management could be provided by the use of molecular assays for gonococcal antimicrobial resistance, particularly in the case of fluoroquinolones [8].

Resistance to all antimicrobial agents has developed in some *N. gonorrhoeae* strains. The global spread of multidrug-resistant *N. gonorrhoeae* is a growing concern, clinical treatment failures with the extended-spectrum cephalosporins (the last option for empirical first-line monotherapy) being recently reported. In many countries, dual antimicrobial therapy (ceftriaxone plus azithromycin) is the recommended first-line empirical treatment [9].

Mycoplasmas, the smallest free-living microorganisms, are another important source of STDs. They belong to the class of *Mollicutes* and were first described in the 1950s, following isolation from a male patient with nongonococcal urethritis [26]. In healthy sexually-active adults urogenital, mycoplasmas are part of the normal commensal flora of the genital tract. *M. hominis*

http://dx.doi.org/10.5772/67871

and *U. urealyticum* are involved in a wide array of infectious diseases, in adults as well as in children. They are suspected of being the causative agents of nongonococcal urethritis, prenatal infections leading to pregnancy complications, infertility, bacterial vaginosis and pelvic inflammatory disease [27].

In contrast to other bacteria, *Mycoplasma* spp. and *Ureaplasma* spp. are not susceptible to penicillins, cephalosporins, vancomycin or rifampicin. Present data suggest that urogenital mycoplasmas are susceptible to agents interfering with protein synthesis, such as the tetracyclines, macrolides, aminoglycosides and chloramphenicol, but also to topoisomerase inhibitors such as the fluoroquinolones. Various mycoplasmal species may pose selective innate resistance to an antibiotic to which other species might be sensitive, for example, all *M. hominis* strains are resistant to erythromycin. Mycoplasma also can develop resistance to antibiotics to which they are usually considered sensitive [28].

As a major cause of nongonococcal urethritis in men, having been directly linked to cervicitis, endometritis, and pelvic inflammatory disease in women, *M. genitalium* holds a distinct position in the spectrum of emerging pathogenic bacteria for humans. It has been isolated from the respiratory tracts and synovial fluid of diseased individuals along with the human pathogen *Mycoplasma pneumoniae* [29–31]. *M. genitalium* is also the smallest known self-replicating cell, with a genome size of only 580 kb, an aspect which imposes severe biosynthetic limitations [32].

Because of extreme difficulties in cultivating *M. genitalium*, many researchers have relied on the results of serological and polymerase chain reaction (PCR) assays to establish links between *M. genitalium* and human disease. Antimicrobial susceptibility testing is consequently rather difficult.

Various investigators noted that resistance to tetracyclines of the mycoplasmas' is on an uphill trend worldwide [33], a Tunisian study reporting resistance rates of 22.7 and 25% among ureaplasmas and *M. hominis*, respectively [34]. Even higher values (45%) were reflected in an American study, in contrast to some European ones suggesting much lower figures [35]. Macrolides and lincosamides are the antibiotics widely used for *U. urealyticum* infections, especially among children and patients allergic to tetracyclines or quinolones. However, widespread macrolide resistance in *U. urealyticum* has been recently reported. In contrast, pristinamycin, a newly promoted macrolide, proved to be ineffective against *U. urealyticum* and *M. hominis* [36]. Still, for sensible patients, for example, premature neonates, one must carefully consider the potential toxicity of the antibiotics in attempting to remove such bacteria [37].

3. Determining the antibiotics resistance of pathogen STD agents

Although resistance to several antibiotics has been reported in vitro, no evidence that *C. trachomatis*—generally treated with a single dose of azithromycin—has developed such resistance were documented [38]. For both chlamydia and syphilis, no internationally agreed methods to assess minimum inhibitory concentrations (MIC) in vitro are standardized [39]. Consequently, assessment of antibiotic resistance relies but on the identification of treatment failures.

The emergence of drug resistance in many pathogenic bacteria compromised severely the therapeutic utility of macrolides. The occurrence of macrolide resistance is yet undocumented for many bacterial pathogens. The various molecular mechanisms inducing bacterial resistance can be collectively characterized as involving drug efflux, drug inactivation or alterations in the drug target site. The probability of developing resistance depends on the types and quantities of drug to which these organisms are exposed [10]. For instance, shortly after erythromycin was introduced in therapy in the 1950s not only was resistance in bacterial pathogens observed, but erythromycin-resistant strains were found to be cross-resistant to all other macrolides and also to the chemically unrelated lincosamide and streptogramin B drugs [40].

Resistance to macrolides of the *C. trachomatis* strains is often associated with mutations in the ribosomal protein genes, particularly in L4 and L22, but also as well as with mutations in the peptidyl transferase region of domain V of the 23S rRNA gene. The later region was demonstrated to play an important role in the interaction of the ribosome with macrolides [10, 11]. Studies on macrolide-resistant *C. trachomatis* clinical isolates led to the first finding of mutations in the peptidyl transferase region of the 23S rRNA gene [12].

Many strains of *C. suis*, a pathogen of pigs, express a stable tetracycline resistance phenotype. This resistance pattern was demonstrated to be associated with a resistance gene, *tet*(C), in the chlamydial chromosome. In tetracycline-resistant *C. suis* strains, four-related genomic islands sharing significant nucleotide sequence identity with resistance plasmids carried by a variety of bacterial species were identified. These genomic islands provided the first examples of horizontally acquired DNA integrated into a natural isolate of chlamydiae [16].

Lateral gene transfer (LGT) is a means allowing *C. trachomatis* to generate variants of enhanced relative fitness, as suggested by the high frequency of between-strain genetic recombinants of *C. trachomatis* among isolates obtained from human STD. Although hampered by the development of the pathogen, experimental investigations on this phenomenon detected in vitro LGT between strains of *C. trachomatis* in vitro. Host cells were simultaneously infected with an ofloxacin-resistant (Ofx^r) mutant of a serovar L1 strain and a rifampin-resistant (Rif^r) mutant of a serovar D strain. DNA sequencing was used to map genetic crossovers. Development occurred in the absence of antibiotics, the progeny being subjected to selection for Ofx^r Rif^r recombinants. Natural DNA transformation is a plausible mechanism, although trans-DNA lengths were previously associated only with conjugation in known microbial LGT systems [41]. Nowadays, LGT studies can be performed with various other *C. trachomatis* combinations to study the mechanisms by which these strains can transfer resistance genes.

As for *C. trachomatis* no reliable laboratory-based gene transfer system was available, in vitro generation of recombinants from antibiotic-resistant strains was used to study the LGT, essential for generating between-strain genomic recombinants of *C. trachomatis* able to facilitate the organism's evolution [42]. For 16 in vitro-derived recombinants of ofloxacin- and rifampin-resistant L1 and D strains, multiple loci were examined and compared with the same sequenced loci among 11 clinical recombinants. Phylogenetics and bioinformatics were used in examining breakpoints and recombination frequency. Without any misclassification, in vitro and clinical isolates clustered perfectly into two groups using Ward's minimum variance based on breakpoint data. *gyrA* (confering ofloxacin resistance) and *rpoB* (confering rifampin resistance), but *trpA* as well, presented significantly more breakpoints among *in vitro* recombinants than among clinical recombinants. Significant selections were evidenced at other loci as well, results indicating that the in vitro model is statistically different from any natural recombination events. Additional genomic studies to determine the responsible factors for selection biases at unexpected loci are needed to clarify whether these are important for LGT approaches in the genetical manipulation of *C. trachomatis* [42].

Several genetic mechanisms are employed by pathogenic microbes for producing variants to counter host defenses. Such a high proportion of urogenital tract isolates presenting amino acid substitutions in the polymorphic *ompA* gene which encodes the major outer membrane protein (MOMP) suggests a role for spontaneous mutation followed by in vivo selection in the bacteria's route to avoid human immune defenses [43, 44]. Humans have diverse B and T cell-mediated responses to MOMP, the occurrence of mutations in the same *ompA* segments of multiple clinical isolates suggesting the protectiveness of at least some of these responses [45]. Evidences also indicate that in vitro LGT may contribute significantly to the origin of enhanced-fitness variants in *C. trachomatis* [41]. The LGT mechanism has not been identified, but various publications investigating possible relationships between in vitro LGT and the high frequency of LGT recombinants in clinical isolates [46] suggested that a certain percentage of the many millions of clinical infections involve more than one strain of *C. trachomatis*. However, a recent study failed to identify mutations in the 23S rRNA genes of resistant mutants selected following enrichment by serial passage in the presence of subinhibitory concentrations of azithromycin [47].

Investigating mutations in the 23S rRNA gene of macrolide-resistant isolates of wild-type *C. trachomatis* obtained from clinical samples and mutant strains selected using subinhibitory concentrations of the macrolides were the objective of another recent study [13] in which a set of resistant clinical isolates of *C. trachomatis* obtained from patients attending the Tianjin Institute of Venerology (Tianjin, China) during 2005 2008 was differentiated into wild-type and mutant strains, the 23S rRNA mutations in the isolates then being identified. Each patient was sampled for only one isolate, the antimicrobial agents examined being erythromycin, azithromycin and josamycin. The in vitro MICs of antibiotics in the isolated clinical strains of *C. trachomatis* were determined. As the MIC values of erythromycin in the eight strains of *C. trachomatis* were higher than the in-blood erythromycin concentration (1 μg/ml) and even higher than the tissue concentration of erythromycin in the urogenital system, these were considered to be wild-type macrolide-resistant strains of *C. trachomatis*.

Following PCR amplification of the 23S rRNA gene of the eight wild-type resistant isolates (exhibiting MIC values above the tissue concentration of the antibiotic present in the urogenital system), no resistance-associated mutations were found at 2057 (*E. Coli* numbering scheme), 2058 or 2059, while only three resistant isolates presented the T2611C mutation. The isolates included the T2611C mutation in the case of two patients with persistent infection, but no mutations were found at 2058. A2057G mutations were found in six mutant isolates, while T2611C mutations were found in 10 mutant isolates. Two mutant resistant isolates presented A2059G mutations, while two of the resistant isolates exhibited no mutations in their 23S rRNA sequences. Medical records indicated no response to azithromycin on behalf of the patients infected with mutant strains. The sensitivity of the wild-type clinical isolates to erythromycin and azithromycin was found to be lower than previously reported [7], explaining the high recurrence rate and treatment failure reported for chlamydial infections. Azithromycin treatment was unsuccessful in eight resistant strains. The wild-type resistant strains presented no mutations in the 23S rRNA, suggesting that other molecular mechanisms were responsible for their resistance. Possible mechanisms underlying drug resistance should be investigated in the future in order to understand the resistance of isolates exhibiting no mutation in the peptidyl transferase region of the 23S rRNA gene. The merits of the Chinese study are that, for the first time, wild-type macrolide-resistant *C. trachomatis* strains have been observed in vitro, and for the first time, A2057G and A2059G mutations in the peptidyl transferase region of the 23S rRNA gene have been found in *C. trachomatis* with selective macrolide resistance [13].

While certain other studies reported that mutations observed in clinical strains were also found in laboratory strains, the reverse was found not to be true likely because rRNA mutations leading to drug resistance in a clinical pathogen often become apparent only when a drug therapy fails to eradicate that pathogen [10].

The first reported cases of clinically significant *C. trachomatis* infection resistant to ofloxacin and azithromycin came from a case study of two patients with *C. trachomatis* infections demonstrating multidrug resistance as they persisted after standard treatment [17]. A commercially available PCR test (Amplicor PCR; Roche Diagnostics, Indianapolis) was employed to detect *C. trachomatis* in urine and urethral/cervical swab specimens. Antimicrobial susceptibility testing on *C. trachomatis* strains isolated from cultured cells originating from urethral/cervical swab specimens was conducted for doxycycline, azithromycin and ofloxacin. In vitro antimicrobial susceptibility testing was also completed.

The MIC (antibiotic concentration level where fluorescent antibody staining failed to highlight typical inclusions after incubation in cell culture) and the minimum chlamydicidal concentration allowing no inclusions on passage in an antibiotic-free medium (MCC) of these antimicrobial agents for these isolates were determined after a subsequent passage of the contents of duplicate unstained wells to a fresh monolayer in antibiotic-free medium. *C. trachomatis* DNA samples prepared from endocervical or urethral swab specimens were amplified by a nested PCR assay, amplicons being purified with a PCR purification kit (Qiagen, Chatsworth, CA). Sequencing was conducted on a model 377 (Perkin Elmer Biosystems, Foster City, CA) automated sequencer using a dRhodamine Terminator Cycle Sequencing kit according to the manufacturer's instructions. Edited sequences were aligned and analyzed with the GCG (Genetic Computer Group, Madison, WI) software package. Genotypes were determined by comparison of resulting sequences with reference *C. trachomatis* omp1 sequences in the GenBank database. Conclusions were that the mechanism responsible for heterotypic resistance in *C. trachomatis* remains unknown, being possible that the observed multidrug resistance to be phenotypic rather than genotypic in nature since the molecular targets of azithromycin, doxycycline and ofloxacin are quite different, and it is highly unlikely that a single or a limited number of gene mutation(s) could be responsible for simultaneous resistance to these diverse agents [13, 48].

Fosfomycin is a broad-spectrum antibiotic that irreversibly binds to the active site of *murA*. Fosfomycin renders the enzyme inactive by forming a covalent adduct with a cysteine residue in the active *murA* site [49]. It was demonstrated [50] that *C. trachomatis* can be resistant to high concentrations of fosfomycin. The genome sequences of *C. trachomatis* contain gene homologues in the peptidoglycan (PG) biosynthesis pathway, including *murA*. The study aimed to demonstrate in vitro activity of *Chlamydia murA* and in vivo activity in *E. coli* and to determine whether *murA* mRNA was expressed in *Chlamydia* bacteria at any time during the chlamydial developmental cycle. The *murA* gene from *C. trachomatis* serovar L2 was cloned and placed under the control of the arabinose-inducible, glucose-repressible ara promoter and transformed into *Escherichia coli*. Findings were that the expression of *C. trachomatis murA* mRNA is cell cycle dependent. Collectively, the data support the notion that *Chlamydia* organisms contain PG and suggest that PG in *Chlamydia* plays a role in development and division. Elucidating the existence of PG in *Chlamydia* spp. is significant for the development of novel antibiotics targeting the chlamydial cell wall.

Another study [51] made use of antibiotic resistance assays and whole genome sequencing to interrogate the hypothesis that two clinical isolates (IU824 and IU888) have acquired certain antibiotic resistance mechanisms (tetracycline efflux mediated by the presence of one of the 28 genes, ribosomal protection via carriage of one of the 10 *tet* genes or one *otr* gene, enzymatic inactivation conferred by one of the three tet genes or mutations in the 16S rRNA gene—identified as lending tetracycline resistance). The genes conferring the first three resistance mechanisms reside in plasmids and insertion elements (horizontally transmissible elements enabling their transfer among a wide range of bacterial species). No tetracycline resistance was evidenced in the two investigated strains, comparisons of the genome and plasmid sequences failing to identify regions able to accommodate horizontally acquired resistance genes.

Inability to introduce stably maintained DNA and perform targeted gene manipulation are long-standing limitations slowing progresses in understanding many of the components associated with the basic biology and pathogenesis of *Chlamydia*. One of these limitations was removed when a novel method for introducing DNA and modificating the native plasmid to allow effective selection of transformants has enabled the development of key molecular tools for studying *Chlamydia*. This transformation system was employed to test the expression of a diverse set of fluorescent proteins and to demonstrate the utility of subcellular localization studies [52]. The *C. trachomatis* developmental cycle incorporates numerous poorly understood processes, but a method for transforming *Chlamydia* has recently enabled the development of essential molecular tools to better study the biology and pathogenesis of these bacteria. Evaluation of green fluorescent protein (GFP) expression demonstrated a tight control of gene regulation by the Tetracycline repressor and anhydrotetracycline. Sensible GFP induction was observed in mid stages of the developmental cycle, data suggesting that metabolic diversity is affecting induction and/or expression during later stages [53].

Previously expensive, laborious and largely inaccessible, routine identification of tetracycline resistance loci became possible through the high-throughput whole-genome sequencing technology that emerged in recent years and enabled the low-cost interrogation of a large number of *C. trachomatis* genomes [54].

N. gonorrhoeae has quite a history in acquiring or developing resistance to antibiotics, including all first-line treatment drugs such as sulphonamides, penicillins, tetracyclines, macrolides, fluoroquinolones or, more recently, cephalosporins. Mechanisms for *N. gonorrhoeae* resistance to antibiotics may involve efflux systems, mutations in the chromosomal targets or acquisition of novel genes by transformation or plasmid-mediated conjugation. Chromosomal mosaic *penA* genes have been recently identified to confer resistance to extended-spectrum cephalosporins including ceftriaxone and cefixime, posing serious threats regarding the development of an untreatable 'superbug' [55].

In bacteria with multiple rRNA (*rrn*) operons, the effect of a beneficial mutation in one operon is likely to be diluted out, thus offering no significant phenotypic advantage. However, amplification of a mutant allele could confer a resistant phenotype, as it occupies the majority of the bacterium's *rrn* operons. While implementation of both of these latter systems requires the acquisition of exogenous genetic material, moderate levels of macrolide resistance have been observed in *N. gonorrhoeae* upon overexpression of an endogenous membrane transport system. The potential risks of resistance developing via modification of endogenous efflux systems such as *mtrRCDE* of *N. gonorrhoeae* or via drug inactivation remain to be assessed [56].

A wide range of plasmids, from the small 4.2 kb plasmid to larger conjugative plasmids promoting their own transfer between strains and carrying determinants of antibiotic resistance, is associated with *N. gonorrhoeae* strains [6].

The use of molecular assays for gonococcal antimicrobial resistance surveillance is particularly suited to the fluoroquinolones, where mutations occur at defined bases in the quinolone resistance-determining regions (QRDRs) of the *N. gonorrhoeae gyrA* and *parC* genes. A real-time PCR (RT-PCR) assay was modified to allow the simultaneous detection of *N.gonorrhoeae* and quantification of the gonococcal susceptibility to ciprofloxacin using clinical samples, the assay being validated with the use of DNA extracted from 40 linked isolates and urethral swabs. The 40 isolates consisted of 11 susceptible isolates, 2 isolates of intermediate susceptibility and 27 resistant isolates. Twenty-four of these linked first-pass urine samples were obtained from men presenting with urethral gonorrhoea [8]. Each gonococcal isolate in the study was reidentified based on Gram staining—a positive oxidase test (Davies Diagnostics, Randburg, South Africa) and coagglutination with the Phadebact® Monoclonal Antibody Test (Pharmacia Diagnostics AB, Uppsala, Sweden). Ciprofloxacin-susceptibility profiles were confirmed using Etest-based MIC assays, while ciprofloxacin susceptibility phenotypes were determined according to Clinical and Laboratory Standards Institute (CLSI) guidelines.

DNA was extracted from an additional 33 first-pass urine specimens collected from patients. A previously published RT-PCR assay differentiating ciprofloxacin susceptible from non-susceptible *N. gonorrhoeae* was modified by inclusion of primers and a probe to detect the gonococcal cytosine-specific DNA methyltransferase gene (*dcmH*) as a positive control target. To generate *gyrA* and *parC* QRDR products for sequencing, amplification of the required sequences was undertaken using isolate-derived DNA. The RT-PCR assay correctly predicted that 15/33 samples were associated with ciprofloxacin-susceptible gonorrhoea, as both *gyrA* and *parC* amplicons were produced. The 18 samples in which *gyrA* amplicons were not generated were correctly predicted to come from patients with gonorrhoea either of intermediate

http://dx.doi.org/10.5772/67871

susceptibility or resistant to ciprofloxacin. The 18/33 samples that produced *parC* amplicons came from 15 patients with ciprofloxacin-susceptible gonorrhoea, 2 patients infected by *N. gonorrhoeae* isolates with intermediate susceptibility to ciprofloxacin and 1 patient with ciprofloxacin-resistant gonorrhoea. DNA extracted from first-pass urine of 15 out of 16 patients with ciprofloxacin-resistant gonorrhoea generated no *gyrA* and *parC* amplicons. The modified RT-PCR assay previously described in this paper offered a faster method for combined *N. gonorrhoeae* diagnosis and determination of ciprofloxacin susceptibility than traditional culture and antimicrobial resistance testing [8].

As infection with *Neisseria gonorrhoeae* is one of the major causes of STDs in Romania, we conducted a study on bacterial *gyrA* and *parC* genes to determine the presence of ciprofloxacin-resistant *N. gonorrhoeae* strains. About 200 subjects from Cluj-Napoca (Romania) suspected of STD infections were enrolled in our study. DNA extracted from first void urine samples was purified and amplified via multiplex PCR for the simultaneous detection of 6 STDs: *Trichomonas vaginalis, M. hominis, M. genitalium, C. trachomatis, N. gonorrhoeae* and *U. urealyticum*. DNA obtained from *N. gonorrhoeae* positive samples was amplified using *gyrA, parC* and *N. gonorrhoeae*-specific *dcmH* gene primers designed using a modified version of previously published sequences [8]. DNA from a *N. gonorrhoeae* ATCC 49226 strain (sensible to ciprofloxacin) served as positive control. The amplicons were separated via gel electrophoresis (**Figure 1**).

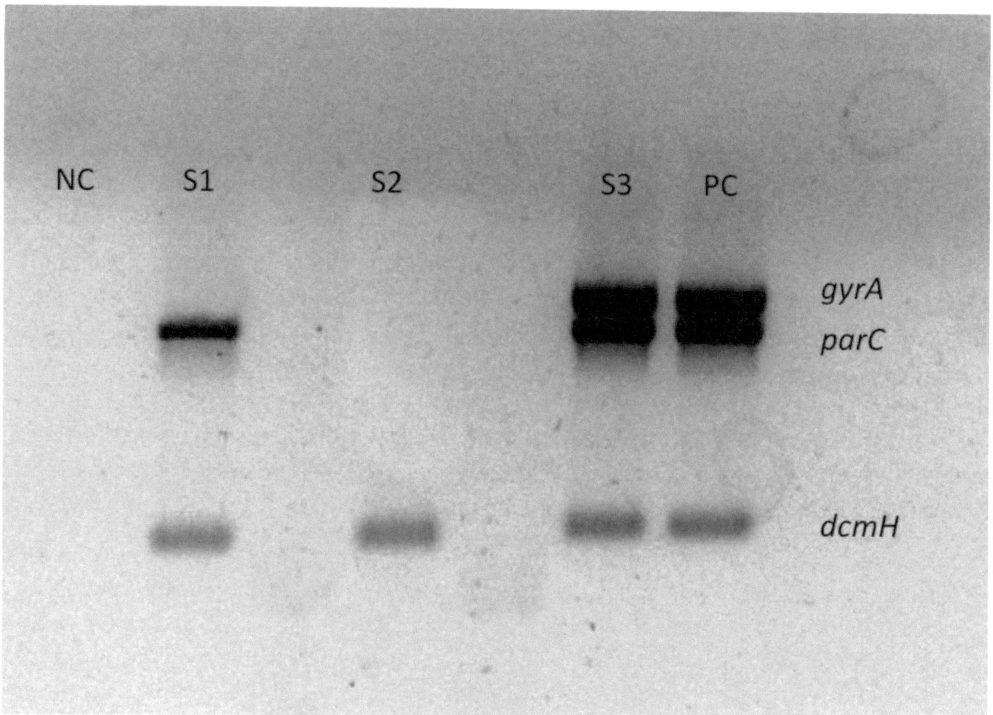

Figure 1. *GyrA, parC* and *dcmH* genes—agarose gel electrophoresis. NC: negative control, S1–S3: samples, PC: positive control.

The *dcmH* gene was amplified in all of the samples collected from the 15 (7.5%) patients diagnosed with gonorrhea, confirming the accuracy of both PCR methods employed in the study. The *gyrA* and *parC* genes were amplified in eight samples accounting for ciprofloxacin-sensitive strains (e.g., Sample S3 in **Figure 1**). The two genes were also present in the positive control sensitive to ciprofloxacin, proving that the primers were correctly designed. The *gyrA* gene was not amplified in six samples presenting only the *parC* and *dcmH* genes, that is, constituting intermediate-resistant strains (e.g., Sample S1 in **Figure 1**). One sample (Sample S2 in **Figure 1**) generated neither *gyrA* nor *parC* amplicons, signifying that the strain was resistant to ciprofloxacin.

This was the first Romanian study on STD pathogens demonstrating the presence of genes providing antibiotic resistance in the *N. gonorrhoeae* strains. The study confirmed the presence of *N. gonorrhoeae* strains resistant to ciprofloxacin in Romania. Most of the strains detected were intermediate-resistant ones.

Elsewhere, mutations at various genes such as 23S rRNA, *mtrR* and *penA* (encoding penicillin-binding protein 2, PB P2) have been identified to associate with chromosomally mediated resistance to azithromycin (AZM) [57] and extended-spectrum cephalosporins [58] in *N. gonorrhoeae*.

A Chinese study combined antimicrobial susceptibility determinations with molecular-based analysis of AZM-resistant in *N. gonorrhoeae* according to the WHO recommendations. Bacterial genome DNA was extracted from each AZM-resistant isolate identified, mutations in the four alleles of the 23S rRNA, *mtrR* and *penA* genes, as well as polymorphisms in *porB* and *tbpB*, being amplified [59, 60].

A high potential for macrolide resistance to occur by mutations in the 23S rRNAs of the bacteria was predicted. The target site for macrolides is the large (50S) subunit of the bacterial ribosome. Many cases of macrolide resistance in clinical strains can be linked to the alteration of specific nucleotides in 23S rRNA within the large ribosomal subunit [10]. Findings were that single base substitutions in rRNA provide macrolide resistance [61].

About 2000 women were tested for vaginal infections. About 1536 of the 2000 vaginal swab samples cultivated in the study [4] were infected with at least one microorganism, a positivity rate of 76.80%. *M. hominis* and *U. urealyticum* combined accounted for <7%. The antimicrobial tests were conducted on Muller agar medium, and antimicrobial activities were evaluated by measuring the diameters of inhibition around the disks. *M. hominis* strains were found to be more resistant than those of *U. urealyticum*.

In a 65-month study conducted on 373 patients attending the Clinic of Dermatovenereology in Belgrade from January 2007 to May 2012, urethral and cervical swabs were analyzed for the presence of *U. urealyticum* and *M. hominis*. *About* 48 (12.9%) of the 373 specimens tested were positive for urogenital mycoplasmas: 37 (77.8%) were found to be positive for *U. urealyticum* and 6 (22.9%) for *M. hominis*, while 5 (10.4%) were positive for both. The test also provided information regarding the density of each organism and its susceptibility to doxycycline, josamycin, ofloxacin, erythromycin, tetracycline, ciprofloxacin, azithromycin, clarithromycin and pristinamycin [62, 63].

http://dx.doi.org/10.5772/67871

Among the 37 isolates of *U. urealyticum*, 34 (94.6%) were sensitive to doxycycline. Furthermore, josamycin and azithromycin (70.3 and 67.6%, respectively) were found to be highly and moderately potent against *U. urealyticum*. *M. hominis* was sensitive to doxycycline (83.3%) and moderately sensitive to ofloxacin (66.7%). *M. hominis* exhibited an absolute (100%) drug resistance rate to erythromycin, tetracycline and clarithromycin, while *U. urealyticum* was highly resistant to clarithromycin (94.6%), tetracycline (86.5%), ciprofloxacin (83.8%) and erythromycin (83.8%). Despite being the most commonly used antibiotic in the treatment of nongonococcal genitourinary infections, doxycycline continues to be the most effective agent for *U. urealyticum* and *M. hominis*. Doxycycline and josamycin were the most potent antibiotics (80%) in mixed infections with both *U. urealyticum* and *M. hominis*. Based on its effectiveness, josamycin could prove to be an alternative treatment for *U. urealyticum* infections. Still, it is believed that it is a matter of time until both mycoplasmas will develop resistance to these new generation antibiotics because of cross-border interaction with resistant agents [62].

Despite the evidence, physicians routinely favor macrolide antibiotics, for example, erythromycin when treating *Ureaplasma* infections in premature neonates [64].

Antibiotic resistance determination of *Ureaplasma* spp. usually requires predetermination of bacterial titer, followed by antibiotic interrogation using a set bacterial input. A method for determining precise MICs and one for screening against multiple antibiotics using breakpoint thresholds were detailed in another study [65] enabling a concurrent determination of the bacterial load in a sample simultaneously with the determination of resistance without prior knowledge of bacterial load. Further PCR and sequence analyses were performed to determine the resistance mechanism upon ureaplasmas grown in commercially available selective medium. An adapted breakpoint analysis was employed to screen for the presence of resistant mutants within the 61 isolates. Confirmation of *Ureaplasma* was determined by amplification of the *Ureaplasma*-specific urease gene (a 430-bp DNA product).

A reliable method to determine antimicrobial susceptibility of *Ureaplasma* isolates without prior knowledge of inoculum size was developed when mutations in bacterial genes previously associated with resistance were investigated for isolates resistant to erythromycin and ciprofloxacin [66]. DNA encoding bacterial 23S rRNA gene, L22 and L4 proteins were amplified and sequenced for macrolide-resistant strains. Gyrase subunits and topoisomerase subunits were sequenced for ciprofloxacin-resistant strains, while *tetM* genes from both tetracycline-susceptible and tetracycline-resistant isolates were also sequenced. More erythromycin-resistant studies are required to determine the prevalence of mutations in the relevant genes associated with resistance.

4. Conclusions

Discovery of antibiotic resistant strains is an important step in choosing the appropriate treatment against bacterial STD after the positive detection of the causative agents. A proper treatment can prevent the spreading of these diseases in the general population. Antimicrobial susceptibility testing proves to be useful in cases when bacteria can be isolated.

The antibiotic resistance of *C. trachomatis* has significantly different characteristics from those of other bacteria. First, because chlamydiae are intracellular pathogens, antimicrobial susceptibility must be determined by their ability to proliferate within a host cell in the presence of variable antibiotic concentrations. Second, in contrast to most bacteria, *C. trachomatis* displays a "heterotypic resistance" in vitro, the chlamydial population containing both susceptible and resistant organisms, only a small proportion of them exerting resistance at any given time [17].

The study of the Chlamydiae biology is complicated by their intracellular development, but knowledge of the recombination mechanism(s) may enable effective genetic manipulations.

Antimicrobial resistance of microorganisms is a quite dynamic phenomenon, highlighting the need for current prevalence and susceptibility data from various different institutions or geographic areas. The ability to use molecular assays for STD antimicrobial resistance surveillance would offer a significant advantage to public health programs and selected patient management.

Drug therapies can result in strains containing mutations that confer the highest resistance becoming prevalent. In contrast, rRNA mutations in laboratory strains are intentionally created in order to evaluate drug interaction mechanisms. The creation of phenotypes with less effective resistance is possible only under controlled laboratory conditions.

The routine use of genomics in clinical settings is becoming a reality. For STDs, understanding the diversity of the circulating strains and their changing over time pattern is a primary requirement. Comparative genomic studies on chlamydia have demonstrated that it undergoes extensive recombination. Genome projects can also help us understand the phenomenon of antibiotic resistance in STD agents, particularly in the cases of *C. trachomatis*, where very little was known about the genetics of these organisms because of a lack of molecular tools and the inability to grow them outside cell cultures or host organisms [14].

In recent year, genomics was asked to clarify a number of clinical aspects. Although genome analysis of bacterial STD was not a main issue, it can undoubtedly reveal previously inaccessible aspects of pathogen biology, including changes that lead to antibiotic resistance thereof. Genetic methods enable the DNA extraction from different biological samples, and both the presence of the bacteria and its resistance to one or more antibiotics can be determined from a single DNA sample. By studying the genes that induce antibiotic resistance and the plasmids that transfer such genes the mechanism that leads to antibiotic resistance can be elucidated.

Author details

Vică Mihaela Laura[1]*, Matei Horea Vladi[1] and Siserman Costel Vasile[2]

*Address all correspondence to: mvica@umfcluj.ro

1 Department of Cell and Molecular Biology, "Iuliu Haţieganu" University of Medicine and Pharmacy, Cluj-Napoca, Romania

2 Department of Legal Medicine, "Iuliu Haţieganu" University of Medicine and Pharmacy, Cluj-Napoca, Romania

References

[1] Nwadioha S., Egesie J.O., Emejuo H., Iheanacho E. A study of female genital swabs in a Nigerian Tertiary Hospital. Asian Pac J Trop Med. 2010;**3**:577–579.

[2] Ekabua J.E., Agan T.U., Iklaki C.U., Ekanew E.I. Adjuncts to case assessment of vaginal discharge syndrome in pregnant women. Asian Pac J Trop Med. 2010;**3**:63–65.

[3] Simpore J., Ilboudo D., Karou D., Pietra V., Granato M., Esposito M., et al. Prevalence of HHV-8 infections associated with HIV, HBV and HCV in pregnant women in Burkina Faso. J Med Sci. 2006;**6**:93–98.

[4] Karou SD, Djigma F, Sagna T., Nadembega C., Zeba M., Kabre A., Anani K., Ouermi D., et al. Antimicrobial resistance of abnormal vaginal discharges microorganisms in Ouagadougou, Burkina Faso. Asian Pac J Trop Biomed. 2012;**2**(4):294–297.

[5] Pawlikowska-Warych M., Śliwa-Dominiak J., Deptuła W. Chlamydial plasmids and bacteriophages. Acta Biochimica Polonia. 2015;**62**(1):1–6. doi:10.18388/abp.2014_764

[6] Snyder L.A.S., Davies J.K., Ryan C.S., Saunders N.J. Comparative overview of the genomic and genetic differences between the pathogenic Neisseria strains and species. Plasmid. 2005;**54**:191–218.

[7] Samra Z., Rosenberg S., Soffer Y., Dan M. In vitro susceptibility of recent clinical isolates of *Chlamydia trachomatis* to macrolides and tetracyclines. Diagn Microbiol Infect Dis. 2001;**39**:177–179.

[8] Magooa M.P., Müller E.E., Gumede L., Lewis D.A. Determination of Neisseria gonorrhoeae susceptibility to ciprofloxacin in clinical specimens from men using a real-time PCR assay. Int J Antimicrob Ag. 2013;**42**:63–67.

[9] Unemo M., Shafer W.M. Antimicrobial resistance in Neisseria gonorrhoeae in the 21st century: past, evolution, and future. Clin Microbiol Rev. 2014;**27**:587–613.

[10] Vester B., Douthwaite S. Macrolide resistance conferred by base substitutions in 23S rRNA. Antimicrob Agents Chemother. 2001;**45**:1–12.

[11] Schlunzen F., Zarivach R., Harms J., Bashan A., Tocilj A., Albrecht R., et al. Structural basis for the interaction of antibiotics with the peptidyl transferase centre in eubacteria. Nature. 2001;**413**:814–821.

[12] Misyurina O.Y., Chipitsyna E.V., Finashutina Y.P., Lazarev V.N., Akopian T.A., Savicheva A.M., et al. Mutations in a 23S rRNA gene of Chlamydia trachomatis associated with resistance to macrolides. Antimicrob Agents Chemother. 2004;**48**(4):1347–1349.

[13] Jiang Y., Zhu H., Yang L.N., Liu Y.J., Hou S.P., Qi M.L., et al. Differences in 23S ribosomal RNA mutations between wild-type and mutant macrolide-resistant Chlamydia trachomatis isolates. Exp Ther Med. 2015;**10**:1189–1193. doi:10.3892/etm.2015.2595

[14] Seth-Smith H.B.M., Thomson N.R. Whole-genome sequencing of bacterial sexually transmitted infections: implications for clinicians. Curr Opin Infect Dis. 2013;**26**:90–98. doi:10.1097/QCO.0b013e32835c2159

[15] Dean D. Pathogenesis of chlamydial ocular infections. In: K. Tasman and E. Jaeger, editors. Duane's Foundations of Clinical Ophthalmology, vol 2. Philadelphia, PA: Lippincott Williams & Wilkins; 2002. pp. 1–20.

[16] Dugan J., Rockey D.D., Jones L., Andersen A.A. Tetracycline resistance in Chlamydia suis mediated by genomic islands inserted into the chlamydial inv-like gene. Antimicrob Agents Chemother. 2004;**48**:3989–3995. doi:10.1128/AAC.48.10.3989–3995.2004

[17] Somani J., Bhullar V.B., Workowski K.A., Farshy C.E., Black C.M. Multiple drug-resistant Chlamydia trachomatis associated with clinical treatment failure. J Infect Dis. 2000;**181**:1421–1427.

[18] Suchland R.J., Geisler W.M., Stamm W.E. Methodologies and cell lines used for antimicrobial susceptibility testing of *Chlamydia* spp. Antimicrob Agents Chemother. 2003;**47**:636–642.

[19] Blythe M.J., Katz B.P., Batteiger B.E., Ganser J.A., Jones R.B. Recurrent genitourinary chlamydial infections in sexually active female adolescents. J Pediatr. 1992;**121**:487–493.

[20] Munday P.E., Thomas B.J., Gilroy C.B., Gilchrist C., Taylor-Robinson D. Infrequent detection of Chlamydia trachomatis in a longitudinal study of women with treated cervical infection. Genitourin Med. 1995;**71**:24–26.

[21] Grayston J.T., Wang S.P. New knowledge of chlamydiae and the diseases they cause. J Infect Dis. 1975;**132**:87–105.

[22] Hogan R.J., Mathews S.A., Mukhopadhyay S., Summersgill J.T., Timms P. Chlamydial persistence: beyond the biphasic paradigm. Infect Immun. 2004;**72**:1843–1855.

[23] Mpiga P., Ravaoarinoro M. Chlamydia trachomatis persistence: an update. Microbiol Res. 2006;**161**:9–19.

[24] Senn L., Hammerschlag M.R., Greub G. Therapeutic approaches to Chlamydia infections. Expert Opin Pharmacother. 2005;**6**:2281–2290.

[25] Horner P. The case for further treatment studies of uncomplicated genital *Chlamydia trachomatis* infection. Sex Transm Infect. 2006;**82**:340–343.

[26] Shepard M.C. The recovery of pleuropneumonia-like organisms from Negro men with and without nongonococcal urethritis. Am J Syph Gonorrhea Vener Dis. 1954;**38**:113–124.

[27] Pararas M.V., Skevaki C.L., Kafetzis D.A. Preterm birth due to maternal infection: causative pathogens and modes of prevention. Eur J Clin Microbiol Infect Dis. 2006;**25**:562–569.

[28] Kenny GE., Cartwright F.D. Susceptibilities of *Mycoplasma hominis, M. pneumoniae,* and *Ureaplasma urealyticum* to GAR-936, dalfopristin, dirithromycin, evernimicin, gatifloxacin, linezolid, moxifloxacin, quinupristin-dalfopristin, and telithromycin compared to

their susceptibilities to reference macrolides, tetracyclines, and quinolones. Antimicrob Agents Chemother. 2001;**45**:2604–2608.

[29] Baseman J.B., Dallo S.F., Tully J.G., Rose D.L. Isolation and characterization of *Mycoplasma genitalium* strains from the human respiratory tract. J Clin Microbiol. 1988;**26**:2266–2269.

[30] Bjornelius E., Lidbrink P., Jensen J.S. Mycoplasma genitalium in non-gonococcal urethritis—a study in Swedish male STD patients. Int J STD AIDS. 2000;**11**:292–296.

[31] Deguchi T., Maeda S. *Mycoplasma genitalium*: another important pathogen of nongonococcal urethritis. J Urol. 2002;**167**:1210–1217.

[32] Tully J., Taylor-Robinson D., Rose D.L., Cole R.M., Bove J.M. Mycoplasma genitalium, a new species from the human urogenital tract. Int J Syst Bacteriol. 1983;**33**:387–396.

[33] Koutsky L.A., Stamm W.E., Brunham R.C., Stevens C.E., Cole B., Hale J., et al. Persistence of Mycoplasma hominis after therapy: Importance of tetracycline resistance and of coexisting vaginal flora. Sex Transm Dis. 1983;**10**:374–381.

[34] Mardassi B.B., Aissani N., Moalla I., Dhahri D., Dridi A., Mlik B. Evidence for the predominance of a single tet(M) gene sequence type in tetracycline-resistant *Ureaplasma parvum* and *Mycoplasma hominis* isolates from Tunisian patients. J Med Microbiol. 2012;**61**:1254–1261.

[35] Farkas B., Ostorházi E., Pónyai K., Tóth B., Adlan E., Párducz L., et al. Frequency and antibiotic resistance of *Ureaplasma urealyticum* and *Mycoplasma hominis* in genital samples of sexually active individuals. Orv Hetil. 2011;**152**:1698–1702.

[36] Karabay O., Topcuoglu A., Kocoglu E., Gurel S., Gurel H., Ince N.K. Prevalence and antibiotic susceptibility of genital *Mycoplasma hominis* and *Ureaplasma urealyticum* in a university hospital in Turkey. Clin Exp Obstet Gynecol. 2006;**33**:36–38.

[37] Waites K.B., Crouse D.T., Cassell G.H. Therapeutic considerations for *Ureaplasma urealyticum* infections in neonates. Clin Infect Dis. 1993;**17**(Suppl. 1):208–214.

[38] Suchland R.J., Sandoz K.M., Jeffrey B.M., Stamm W.E., Rockey D.D. Horizontal transfer of tetracycline resistance among *Chlamydia* spp. in vitro. Antimicrob Agents Chemother. 2009;**53**:4604–4611.

[39] Norris S.J., Cox D.L., Weinstock G.M. Biology of *Treponema pallidum*: correlation of functional activities with genome sequence data. J Mol Microbiol Biotechnol. 2001;**3**:37–62.

[40] Leclercq R., Courvalin P. Bacterial resistance to macrolide, lincosamide, and streptogramin antibiotics by target modification. Antimicrob Agents Chemother. 1991;**35**:1267–1272.

[41] Robert DeMars R., Weinfurter J. Interstrain gene transfer in *Chlamydia trachomatis* in vitro: mechanism and significance. J Bacteriol. 2008;**190**(5):1605–1614.

[42] Srinivasan T., Bruno W.J., Wan R., Yen A., Duong J., Deborah Dean D. In Vitro Recombinants of antibiotic-resistant *Chlamydia trachomatis* strains have statistically more breakpoints

than clinical recombinants for the same sequenced loci and exhibit selection at unexpected loci. J Bacteriol. 2012;**194**(3):617–626. doi:10.1128/JB.06268-11

[43] Millman K., Black C.M., Johnson R.E., Stamm W.E., Jones R.B., Hook E.W., et al. Population based genetic and evolutionary analysis of *Chlamydia trachomatis* urogenital strain variation in the United States. J Bacteriol. 2004;**186**:2457–2465.

[44] Ortiz L., Angevine M., Kim S.K., Watkins D., DeMars R. T-cell epitopes in variable segments of *Chlamydia trachomatis* major outer membrane protein elicit serovar-specific immune responses in infected humans. Infect Immun. 2000;**68**:1719–1723.

[45] Dean D., Millman K. Molecular and mutation trends analyses of omp1 alleles for serovar E of *Chlamydia trachomatis*. Implications for the immunopathogenesis of disease. J Clin Investig. 1997;**99**:475–483.

[46] Sayada C., Denamur E, Grandchamp B., Orfila J., Elion J. Denaturing gradient gel electrophoresis analysis for the detection of point mutations in the *Chlamydia trachomatis* major outer-membrane protein gene. J Med Microbiol. 1995;**43**:14–25.

[47] Binet R., Maurelli A.T. Frequency of development and associated physiological cost of azithromycin resistance in *Chlamydia psittaci* 6BC and C. trachomatis L2. Antimicrob Agents Chemother. 2007;**51**:4267–4275.

[48] Rice R.J., Bhullar V., Mitchell S.H., Bullard J., Knapp J.S. Susceptibilities of *Chlamydia trachomatis* isolates causing ncomplicated female genital tract infections and pelvic inflammatory disease. Antimicrob Agents Chemother. 1995;**39**:760–762.

[49] Kahan F.M., Kahan J.S., Cassidy P.J., Kropp H. The mechanism of action of fosfomycin (phosphonomycin). Ann N Y Acad Sci. 1974;**235**:364–386.

[50] McCoy A.J., Sandlin R.C., Maurelli A.T. In vitro and in vivo functional activity of Chlamydia MurA, a UDP-N-Acetylglucosamine enolpyruvyl transferase involved in peptidoglycan synthesis and fosfomycin resistance. J Bacteriol. 2003;**185**(4):1218–1228.

[51] O'Neill C.E., Seth-Smith H.M.B., Van Der Pol B., Harris S.R., Thomson N.R., Cutcliffe L.T., et al. *Chlamydia trachomatis* clinical isolates identified as tetracycline resistant do not exhibit resistance in vitro: whole-genome sequencing reveals a mutation in porB but no evidence for tetracycline resistance genes. Microbiology. 2013;**159**:748–756.

[52] Agaisse H., Derré I. A C. trachomatis cloning vector and the generation of *C. trachomatis* strains expressing fluorescent proteins under the control of a *C. trachomatis* promoter. PLoS One. 2013;**8**:e57090. doi:10.1371/journal.pone.0057090

[53] Wickstrum J., Sammons L.R., Restivo K.N., Hefty P.S. Gene expression in *Chlamydia trachomatis* using the Tet system. PLoS One. 2013;**8**(10):e76743.

[54] Harris S.R., Clarke I.N., Seth-Smith H.M., Solomon A.W., Cutcliffe L.T., Marsh P., et al. Whole-genome analysis of diverse *Chlamydia trachomatis* strains identifies phylogenetic relationships masked by current clinical typing. Nat Genet. 2012;**44**(S1):413–419.

[55] Unemo M., Shafer W.M. Antibiotic resistance in *Neisseria gonorrhoeae*: origin, evolution and lessons learned for the future. Ann N Y Acad Sci. 2011;**1230**:19–28.

[56] Hagman K.E., Pan W., Spratt B.G., Balthazar J.T., Judd R.C., Shafer W.M. Resistance of Neisseria gonorrhoeae to antimicrobial hydrophobic agents is modulated by the mtrRCDE efflux system. Microbiology. 1995;**141**:611–622.

[57] Chisholm S.A., Dave J., Ison C.A. High-level azithromycin resistance occurs in *Neisseria gonorrhoeae* as a result of a single point mutation in the 23S rRNA genes. Antimicrob Agents Chemother. 2010;**54**(9):3812–3816.

[58] Liao M., Gu W.M., Yang Y., Dillon J.A. Analysis of mutations in multiple loci of *Neisseria gonorrhoeae* isolates reveals effects of PIB, PBP2 and MtrR on reduced susceptibility to ceftriaxone. J Antimicrob Chemother. 2011;**66**(5):1016–1023.

[59] Liang J.Y., Cao W.L., Li X.D., Bi C., Yang R.D., Liang Y.H., et al. Azithromycin-resistant *Neisseria gonorrhoeae* isolates in Guangzhou, China (2009–2013): coevolution with decreased susceptibilities to ceftriaxone and genetic characteristics. BMC Infect Dis. 2016;**16**:152–159. doi:10.1186/s12879-016-1469-3

[60] World Health Organization. Global action plan to control the spread and impact of antimicrobial resistance in *Neisseria gonorrhoeae*. 2012.[Internet]. 2012. Available from: http://www.who.int/reproductivehealth/publications/rtis/9789241503501/en/ [Accessed: 21 October 2016].

[61] Sor F., Fukuhara H. Identification of two erythromycin resistance mutations in the mitochondrial gene coding for the large ribosomal RNA in yeast. Nucleic Acids Res. 1982;**10**:6571–6577.

[62] Skiljevic D., Mirkov D., Vukicevic J. Prevalence and antibiotic susceptibility of *Mycoplasma hominis* and *Ureaplasma urealyticum* in genital samples collected over 6 years at a Serbian university hospital. Indian J Dermatol Venereol Leprol. 2016;**82**:37–41.

[63] Clegg A., Passey M., Yoannes M., Michael A. High rates of genital Mycoplasma infection in the highlands of Papua New Guinea determined both by culture and by a commercial detection kit. J Clin Microbiol. 1997;**35**:197–200.

[64] Matlow A., Th'ng C., Kovach D., Quinn P., Dunn M., Wang E. Susceptibilities of neonatal respiratory isolates of *Ureaplasma urealyticum* to antimicrobial agents. Antimicrob Agents Chemother. 1998;**42**:1290–1292.

[65] Beeton M.L., Chalker V.J., Maxwell N.C., Kotecha S., Spiller O.B. Concurrent titration and determination of antibiotic resistance in *Ureaplasma* species with identification of novel point mutations in genes associated with resistance. Antimicrob Agents Chemother. 2009;**53**(5):2020–2027.

[66] Bebear C.M., Renaudin H., Charron A., Gruson D., Lefrancois M., Bebear C. In vitro activity of trovafloxacin compared to those of five antimicrobials against mycoplasmas including *Mycoplasma hominis* and *Ureaplasma urealyticum* fluoroquinolone-resistant isolates that have been genetically characterized. Antimicrob Agents Chemother. 2000;**44**:2557–2560.